THE ENGENNER

THE SECRET OF **SONY** TECHNOLOGY

NOBUTOSHI KIHARA

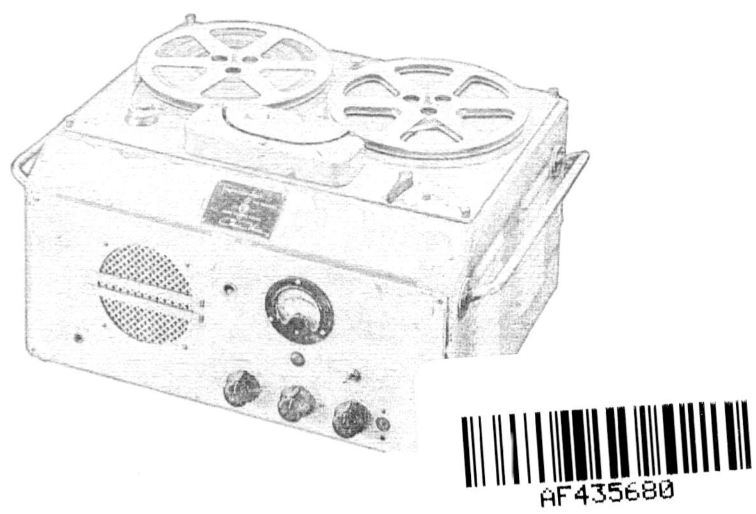

FIELD ARCHIVE INC.
TOKYO, JAPAN

FIELD ARCHIVE
TOKYO, JAPAN

Cover design : Field Archive Inc.
Interior design : Tomomi Kihara (Field Archive Inc.)
Typography : Hiroaki Kurokawa (Field Archive Inc.)
English translation : Yu Matsumoto
English Proofreader : Malcolm Hendricks

Originally published in Japan by
Field Archive Inc., 2019

Am I able to clearly show what goals and objectives should be
passed on to future generations?
"yes, I can"

- Nobutoshi Kihara

CONTENTS

THE
ENGENNER

INTRODUCTION

The reason I thought about writing this book in the first place is that, when I looked back at the history of technology in the past half-century and pondered about what was born from it, I reached my own conclusion.

If someone was to ask me, "Did you do something good for the world? " I can answer "yes" with confidence.

Throughout history, in order to build a happier and a more peaceful civilized society, people shared their wisdom with one another to advance one step at a time. If that wisdom were to disappear in just one generation, the next-generation must start the same cycle all over again, and there will be no accumulation of knowledge or advancement. Human society would be driven solely by animal-like instinct.

But somewhere down the line, people learned to communicate with words, learned to paint, invented writing, engraved words in stones to pass their history along to future generations, used papyruses and paper which enabled communication with people from afar, and then invented ways to mass print on paper —thus began an era of civilization in which massive amounts of books could be disseminated.

The driving force behind the development of civilization is its methods of communication and recording said communication for future generations.

In modern times, people have become capable of transmitting information to broader areas thanks to the emergence of the telephone which enabled instant conversation, and images could be saved and accumulated by taking pictures using film. Documents could be widely distributed using fax machines, and movies could be viewed by many at cinemas. Now, with rapid development, phonographic discs have been replaced by tape recorders, CDs and MDs that use optical recording technology, as well as television and video recording. In addition, the ability to store data onto vast memory-storage devices using computers is rapidly improving.

Advancements in recording technology – due to the development of electronics in the 50 years after World War II – have been dramatic; magnetic recording in particular was at the top of these advancements. Audio, image, movie – all could now be saved and passed on to the next generation.

I spent half a century diligently developing the magnetic recording technology that became one of the drivers of the development of civilization, allowing us to pass on the accumulated wisdom of our forerunners to future generations. All I did was just work hard, but now that I think back on those days, I feel that my work was something that a great deal of good for society.

Am I able to clearly show what goals and objectives should be passed on to future generations? I would very much like to answer "yes, I can" to that question too.

It has been several years since I started to think about having to pass on all of the technical knowledge that I accumulated during my feverish years of work, and finally, I have managed to do just that. This book is the result of that work.

I believe that what I thought and what I created by the sweat of my brow can only be written by me, myself. I have been interviewed on countless occasions and appeared in newspapers and magazines, but those are all me seen from the third person, and my opinions are all filtered in one way or another.

For this reason, I wrote this book to record my own true thoughts, experiences, and happenings, as well as episodes that I have never told.

That is the desire that I have put into this book. I believe that engineers experience their greatest pleasure only when they feel the happiness or the sense of satisfaction that the technology that they created did some good for the advancement of civilization. Of course, if the readers of this book, not just engineers, are able to empathize with any part of this book, put it to practice and find pleasure in it, that would be more than I would hope for.

Spring, 1997
- Nobutoshi Kihara

CHAPTER ONE

LUCK IS DRAWN TO STRONG WILL

How I took my first step to "SONY of the World"

1. Development is the Exploration of the Possibility to Make a Dream Come True

THE GODDESS OF DESTINY SMILED

"I went to listen to the tape recorder, and it sounded much better than the wire (recorder). Are you interested in doing a research on tape recorders? "

It was early June in 1949 when Mr.Masaru Ibuka, the founder of Sony, summoned me and asked me this question. That was enough to ring a bell inside my head.

"Yes, I'll do it."

I answered right away, and that was when my big change from wires recording devices that I had been developing to the research of tapes started.

In retrospect, I think that the Goddess of destiny smiled upon me that day. And I believe that this was the day that the "luck" of Tokyo Tsushin Kogyo K.K. (Tokyo Telecommunications Engineering Corporation, the predecessor of Sony) started, as well.

I am by no means a fatalist, but I firmly believe that as long as you have a strong will to keep steadily advance toward your goal, you can find a solid path to break new ground, however high the barriers that may appear in front of you.

In this book, I would like to tell the story of how we navigated each fork in the road of destiny during my long, long history with Sony that started on this day, through to the world of technology that I navigated.

What Ibuka-san told me was that a brown-surfaced tape about 6mm wide was wound onto a reel, and sound was recorded onto it and played back by winding the tape. That was all the information given to me at that time, but I took the hint. I figured that there must be some kind of magnetic powder applied to the surface of the tape.

Ibuka-san showed me with his fingers that "the reel was rotating at about this much speed," so I was able to guess that the running speed of the tape was about 20 to 30cm per second.

The Clue Was on the Bookshelf

First and foremost, you can't do any experimenting without the tape, so I started off by searching for the magnetic powder. Right away, Ibuka-san brought me a chunk of something that looked chalky and black, saying, "how about this OP magnet? " and left it on my work table, with the suggestion to "try grinding it in a mortar."

I knew about the OP magnet too since it was used for the speakers' magnetic fields, so I thought, "my hands will get all black doing this, but oh well, I'll have a mortar prepared for me and start grinding." While waiting for the ingredients to arrive, I thought about studying magnets in the library, but checked the company's bookshelf first, since it was much closer than the library. There, a thin paperback-sized book about magnets written by Kotaro Honda (Former provost of Tohoku University and inventor of KS steel) caught my eyes.

Among the description of various magnets, a mere three sentences mentioned "retort the yellow powder of ferric oxalate and remove water and carbon dioxide, and the remaining brown powder is 'gamma hematite' (Fe_2O_3), which produces a bar magnet when tamped down in a glass tube." I found this worth experimenting with, so I asked to purchase some of this ferric oxalate. Then, Mr. Akio Morita (Sony's co-founder) got up from his chair and said, "I know a wholesaler of chemicals in Kanda, so I'll go with you and find it." We took the commuter train Yamanote line to Kanda, and searched several stores until, at last, we found the yellow ferric oxalate powder and purchased two bottles.

It was only a few years after the war, and Tokyo was still in the ruins. To find a very special chemical like ferric oxalate was nothing but sheer good fortune, given that this chemical had enabled us to discover that the magnetic tape could be developed in a short amount of time. If we hadn't found it, we probably would have been forced to take a big detour, and might have even gotten lost along the way. I suppose we would have developed the tape in the end either way, but a huge amount of time and effort would have been spent in vain.

Quite recently, a TV station requested to film a reenactment of the scene where the tape was developed, so I ordered some ferric oxalate to show the TV crew how the powder transforms when heated in a pan, but I was told that ferric oxalate could no longer be found in Japan. But luckily, a vendor in the industry told me that there was some imported from the States, so finally, I could get a hold of just a small amount for a very high price. After nearly half a century, in the era when anything can be bought by money, I had a hard time finding ferric oxalate. The fact that I could find it back then led me to strongly believe that the Goddess of destiny had smiled upon me.

Now, looking at the tiny bottle of ferric oxalate in my hand and recalling the book in the bookshelf inviting me to "check this chemical out!" and the quick response of Morita-san to run to Kanda to find it, I think that it was much more than just luck, but rather, a natural consequence of all the effort and the earnest attitude of everyone toward their work, which fills me with deep emotions anew.

The Production of "Magnetic Powder"

Ferric oxalate is a fine yellow-colored powder, so fine that it would disperse under one breath. I think we had a microscope in the office, but rather than seeing how small the molecules were, I was so eager to produce magnetic powder that I couldn't stop myself from running to where we had gas, in this case, the kitchen.

I asked Irinami-san, who always cooked for us in the kitchen, to let me use a pan and spatula to stir-fry the ferric oxalate over the fire, shuffling it with the spatula like roasting wheat. At first, the yellow gradually turned brown.

I thought it was the exact same reaction that the book described, but in a few moments, it started to turn black (magnetite Fe_3O_4). Then, the reaction sped up, catching fire from the rim, and in the end, the color changed to rouge (red oxide).

I found out later that the brown was the gamma hematite, which would become the mainstream ingredient for tape production.Magnetite was used for a short while too, but soon fell into disuse.

As I continued to stir-fry the ferric oxalate, I finally got the knack of it, and learned to stop the reaction abruptly by putting the chemical in a bowl of water just before it turned from brown to black.Such craftsmanship to stir-fry powder !

I hadn't forgotten about the OP magnet, but the mortar arrived in the meanwhile, so I got to grinding the powder, and was amazed to find how hard the chemical was. It took a very long time to grind it into fine powder, so I finally gave up, and mixed it with rubber cement while it was still coarse, and pasted it onto a tape with a paintbrush. At first, I had been using cooked rice mixed in water to produce starch, and then mixed it with the OP powder to paste it onto the tape. At the time, I was so eager to know if sound could be recorded on the tape that I was experimenting with anything that I could get my hands on that functioned as glue.

As might be expected, rice starch turned all crispy when dry, and since it lacked flexibility, it cracked up and didn't work. The next thing I came up with was gum Arabic that was used in the office. I could hardly wait for it to dry, and stretched a 10cm-long OP tape to move it in front of the head. I heard a sound. But it was a rasping sound.

The much-expected first tape seemed to be a failure, so I stopped the experiment right away, and shifted to a new idea, the "rust-collection project." The idea was that if you collected rust that forms on the steel cable that was used in wire recorders, maybe it could be used as magnetic powder. This project failed too, since we couldn't collect as much rust as I had hope, not enough to apply it onto the tape, so I gave up on this plan.

All this might sound like a few days' work, but in fact it was all done in just one day.

I eventually found out that the OP magnet was not a type of oxidized iron but rather a type of ferromagnetic alloy powder that was hardened. In principle, OP can be used for magnetic recording, but considering the state of technology at the time, oxidized iron was sufficient for tape recorders. The reason I could produce only a rasping sound was because the magnetic field produced by the very strong magnetism could not be eliminated with the standard eraser head at the time. As a matter of course, recording was not possible even with the application of a high-frequency bias.

At the time, "stainless" referred to non-magnetic objects, which means that its magnetism was so weak that it barely stuck to a magnet.

This magnetic alloy that made up the OP magnet had a very attractive feature as an ingredient for magnetic tapes, and as we entered the era of video tapes, it came to be a mainstream ingredient for magnetic tapes, which is quite funny to think back to now.

I started off with stir-frying the powder in a frying pan, but then Morita-san bought me an electric furnace and thermometer so that I could cook the powder in a more stabilized environment. I remember it was very hot that summer, but that might be because I was constantly standing in front of the frying pan or the furnace. One very hot day, I walked away from the furnace just for a little while to cool off, and the expensive thermometer melted! I apologized profusely to Morita-san, but that was the only time I made such a mistake. It really must have been hot that summer – and it was only the end of June.

Handmade Magnetic Head

Two things you cannot do without in order to produce sound from a tape is the head and the record/play circuit. If you are in a big hurry to run an experiment, this is all you need to produce sound. You don't need any machines.

You can record/play sound by pulling the tape with your hand and rubbing it in front of the head. Even by using such crude methods, I was rushing to evaluate sample after sample.

At my workplace, I was not in the mood to patiently measure results. Ibuka-san and Morita-san would pop their heads in every once in a while and ask, "How about this ingredient? ", "What about this paintbrush? ", "This paper? ". Or, "This glue? " I was grateful, but I had narrowed down the ingredient to ferric oxalate and was too busy running cut-and-try experiments, trying to somehow produce sound.

It was obvious that the record circuit experimented on wire could be reused, so next, I had to create the head that would prove indispensable for the experiment. But I had never seen or heard of a head for a tape. I used my imagination and analogized from the wire recording head.

I cut a permalloy board in 6mm widths, and bent two of those to create a gap. The magnetic property of permalloy retrogrades extremely after mechanical distortion, so after a process like the one I did, it must be sent out to be annealed. Creating the head turned out to be a very burdensome task.

The Whetstone Skill that My Father Taught Me Came in Handy

When I started on the development of wires, prototyping of the head had been completed. So we had a stock of permalloy boards, and we also had experience in winding techniques and coil impedance (a coefficient that represents how difficult it is for the alternating current to flow) measurement. The shape might be different, but the principle is the same, so I didn't think that there would be any problem.

However, I wasn't sure how much gap length (opening) the head required. Considering the ratio of the speed per second of the wire and the tape, I assumed that one fifth or a quarter should be sufficient, and decided to press-roll a beryllium copper plate to 0.02mm using a pressing machine and place it between the gap.

The quality of a head is determined most by the accuracy of the gap. The plate thickness of the wire head is about 1mm, and the gap is about 0.1mm, so I didn't have to worry too much about the accuracy, but with a tape whose width is as wide as 6mm, creating a uniform gap of 0.02mm became a problem.

At the time, surface grinding and lens polishing technology was quite advanced, so I think it would have been possible to fabricate the head by drawing a design and ordering it from one of our subcontractors, but there was no time for anything like that. I decided to handcraft it myself.

I had an annealed 1mm-thick permalloy board which I cut into 6mm widths and bent into a U-shape with sharp corners, and set two of them together to create a looped magnetic circuit with a gap. It could be wound later, but first, I had to work on the gap. If the line is not linear for the whole 6mm length, the 0.02mm gap length cannot be maintained, and would become an X-shape with both ends bent outwards or an O-shape with the center expanded. Even if the margin of error was 10%, its accuracy must be held at two thousandths of a millimeter. Supposing that was possible, how was I to measure the accuracy?

When I was in college, I worked as an assistant at Kihara Casting Factory run by my aunt's family. I used to measure the thickness of aluminum and copper plates with a micrometer and distribute them to the operators. One time, wondering what a hundredth of a millimeter was like, I measured anything that seemed thin. This was because a hundredth of a millimeter was the thinnest that the micrometer could measure.

While I was playing around, I held up the micrometer in the air to check if I could see a gap when the scale was set at a hundredth of a millimeter without any object held in it, and found that I could see a gap wider than I had imagined.

Recalling that experience, I borrowed the micrometer that the finishing operators Kei Yamada and Shintaro Kimura kept in the workshop that I always worked in and took a closer look. I could certainly see light coming through the hundredth of a millimeter gap, but when I tightened the gap further, the light suddenly vanished. Okay, this is it. Determined to apply this method, I immediately moved on to processing the gap.

Gap processing is essentially a sharpening process using whetstones, so the baseline of the accuracy depends on the surface accuracy of the whetstone.

My father had taught me how to use the whetstone when I was a child, so it was a piece of cake to sharpen kitchen knives and planes, and I even knew how to listen to the sound of the whetstone and of metal rubbing against metal to maintain a constant angle when sharpening the blade. I suppose this is something called "know-how." Another thing that my father taught me was how to level out the surface of the whetstone.

After being used for a long time, the whetstone sags especially in the center part, so you need to take care of it quite frequently. To do so, you rub two similar-sized whetstones together and level out the surface. You need to rinse it down with water repeatedly to check where the whetstones touch each other, and continue to rub them together until the surface becomes evenly flat.

Applying this method, I hand-polished the gap section of the head using the whetstone that I had borrowed from the finishing operator. I put the two U-shaped heads together and held them under the light, and saw that the center section did not let light through but the end sections looked brighter. This means that the gap was polished in an X-shape. Once I knew that, the rest was easy. Imagining that I was Hidari Jingoro, the legendary Japanese sculptor of the Edo period, I stabilized my arms so that the heads would not tilt left or right, taking every precaution to not change the rubbing sound, I polished them lightly just a few more times. I put the heads together again, and this time, there was no light coming through. I suppose this was due to little more than my own hunch.

Next came the winding step. A 0.1mm enamel wire had to be wound about 1,000 times one way, so I borrowed the manual winding machine from Sadashi Kurokami and wound the wire onto the head's bobbin (the frame onto which the wire is wound).

The head production was finished, but the final challenge was the grafting of the gap sections. Since it is U-shaped, there will be two gaps back and front. The front gap is the important side where the tape touches, but the back gap is not as important, so I soldered that section first. If I had made a proper jig beforehand, it would have been much easier, but I was in a rush to produce a head for experimenting, so I asked Mr.Kimura, the finishing operator, to hold the parts down with pliers and tweezers. Once the back gap section was grafted, I tucked the thin plate of beryllium copper in the front gap, soldered it down, then polished the traveling surface of the tape, and the head was complete.

With such crude handcrafting and soldering, I was worried that the permalloy quality might be too deteriorated for use, but the OP magnet tapes and rust tapes were able to produce a rasping sound, so the gap was capable of picking up magnetism. This is how I made the head for experimenting.

The Tape Produced a Sound

As for the measuring equipment, I spent less than half a day making it when the decision was made to assign me to the development of the tape. I made an endless tape, fitted a rubber roller with about a 5cm diameter in the center of the phono-motor turntable for a 78rpm radio phonograph, then screwed on the table for the head and had an amplifier ready too. I had not expected much from the OP magnet tape, but was still disappointed to learn that I couldn't produce a good enough sound. The rust tape was the same.

Finally, I cooked enough ferric oxalate powder to apply to the tape, so I had planned on making about 3m of tape for recording and playing. But rubber the cement had been too thickly applied with the paintbrush, and the surface was not flat. Ibuka-san brought me a brush made from raccoon chest-hair, but that too could not produce the desired result.

Still, I was able to make a material of about 10cm in length, so I held it with my hands just like when I experimented with the OP tape, and moved it in front of the head to record and play. It barely produced any kind of noise, and the recorded "ah, ah" came out like "ua, ua".

I thought that it was a failure, but the difference from the OP tape was that something more than just noise had been recorded. I had a feeling that if I could apply the powder more evenly onto the tape, the sound would be better.

The next day, someone told me that the new plant on the top of the hill had been completed, and that you could go inside. So, I climbed up the brand-new stairs and took a look inside. I was looking around, impressed at how big the place was, and saw a painter using a spray-gun to spray paint onto a wall on the far side. I instantly had an idea. What a bubblehead I had been to try to paint with a brush! I went to the painter right away to ask about the paint, the binder, the blending and the spraying process, and even managed to negotiate borrowing the spray-paint machine.

That evening, once the painting work of the new plant was finished, I put the black-brown liquid consisting of ferric oxalate magnetic powder and clear lacquer in the spray gun, and sprayed it onto the tape. It was a great success, but I never imagined that this spray-gun painting would cause a big problem. I had managed to paint the floor of the brand-new plant with black powder. The plant's general manager, Akira Higuchi (former vice president) scolded me and said, "what are you going to do about this? !" and I felt very sorry, but nothing came of it, and the floor was fixed in no time. I suppose the painter worked to restore it.

Thanks to the spray paint, the tape replayed my voice,

"Honjitsu wa Seiren Nari (It is a fine today)."

It was a much clearer sound than the wire, and the very first tape-recorded sound in Japan. The tape was about 3m long, so I set it in the hand-made endless tape recorder/player. Then, an excellent quality sound was recorded and played. It was so accurate that there was almost no need to adjust the voltage value of the high-frequency bias for recording or the head impedance.

I reported this to Ibuka-san and Morita-san right away, and had them listen to it. The two men were extremely pleased by this result.

Picking up clues one after another to achieve one goal from zero, so to speak, since we knew nothing about tape development at the time, gave me a sense of satisfaction and achievement that almost made me want to shout "I did it! "

tape recorder experiment kit

Similarly, the story of Sony Corporation's performance improvement continued on endlessly hereafter, but at each stage, whenever we achieved something of the highest quality, we called it the "God's head" or the "God's tape" out of respect, and handled them with care as a benchmark head or tape to compare performance.

Development of the Wire Recorder

Let's go back in time a little. A year earlier, in autumn of 1948, my telegraph key facsimile project (discussed later) was nearing completion, so I was assigned to another new product development. I think that at the time, Ibuka-san and Morita-san had wanted to develop a product that was more for the public. Mr.Masanobu Tada of Nippon Denki (now NEC) happened to bring a steel-wire magnetic recorder (wire recorder), and it was determined that this would be our next product.

So I was summoned immediately and was given the machine with an order to "study this wire recorder."

I am always very curious about anything new, so excitedly, I turned on the power of this wire recorder and operated it according to the manual.

I had been told that it was a military-grade machine, and it was accordingly very sturdy, with a very heavy and stiff transition handle, so it was highly unlikely that women or children could handle it. And since the magnetic recording wire had been oiled to prevent rust, the machine was all greasy inside.

Nevertheless, it was a valuable object of interest to me, and I had fallen head over heels about the complexity of the mechanism, the amusing movement, and the novelty of the recording electric circuit. When I inspected the mechanism, no questions came up, and looking at the schematic and tracing the movement along the wire connection, I could understand it fully.

Tracing the schematic, I recorded the voltage of major points, the wave profile of the movement and the amplification.

This was completed in a few days, but what interested me most was that a high-frequency bias method (a method in which sound is recorded along with a high frequency) was used to record onto the steel wire.

Ever since I was a junior high school student, I used to read books like "MUSEN TO JIKKEN (The Radio Experimenter's Magazine)" and "KODOMO NO KAGAKU (Children's Science)", so I had basic knowledge of magnetic recording. I knew that in Germany and England, there were machines that could record sound onto a band of steel using the principle of magnetism, which applied a biased magnetic field to play a distortion-free sound. This was probably the direct current magnetic field, but I had no knowledge of a method that applied other high-frequency biases.

I questioned Tada-san about this one day, and he told me that "it is a method called the alternate current bias or the high-frequency bias, and it records with much higher sensitivity than the direct current bias with more output, so it produces less noise or distortion when played. It is the latest method."

I was all Alone

Looking back now, I don't remember ever thinking "oh, this is a big problem" when I was handed the wire recorder and ordered to study it. Rather, I was overjoyed and voluntarily analyzed the mechanism one after another.

No one but Tada-san knew about magnetic recording back then, so if I was faced with something I didn't understand, I had to figure it out on my own. This was because he was not an employee of Totsuko (Tokyo Tsushin Kogyo) at the time, and there weren't a lot of chances to meet him. Everybody else had projects and was working so hard, I wouldn't dare disturb them. The thought of relying on others hardly crossed my mind.

The senior executives especially were putting an unimaginable amount of work and effort in managing the company. After having more recently read "Genryu," Sony's 40-year company history, I can see that there was a long history of all-out effort. When I was studying the wire recorder, Morita-san had been importing a wire recorder kit made by Webster Chicago through an American friend of his.

The moment he decided that this was to be the next project for the company, he searched high and low to gather information, and was capable enough to take the initiative in getting hold of what was necessary.

Of course, he never mentioned a word about that to me until one day, he suddenly brought me the wire recorder kit and told me to set it up and operate it. Did he think that I was capable of doing that, or did he just think that I had enough free time to play around with the kit? I think it was the latter. Anyhow, I worked all alone and relied on my gut feeling to build it up.

At the time, I did not belong to any department, and was working totally alone. I don't remember having a direct superior to whom I reported or consulted with, or who gave me my orders. All I remember is Ibuka-san or Morita-san coming directly to me, and reporting directly to them "the machine worked" or "I made what you asked for."

Webster Chicago's wire recorder kit
(photo from Masanobu Tada
'Magnetic Recorder', Ohmsha, 1953)

Accumulation of Know-how

All the kit contained was the mechanical parts to drive the wire, so the electric components had to be hand-made. The circuit was a simple one along the lines of a five-tube super, so I went to Kanda to buy a chassis, switch, transformer, vacuum tube and resistance capacitor. I assembled the parts, wired them and adjusted the machine, but the frequency of the high-frequency transmitter and the power did not seem to be aligned, so I tried rewinding the transmitting coil several times and was finally able to record and play. After that, the performance steadily improved, but this was the first time that I became painfully aware of how difficult circuit wiring was.

In addition, recorders had unimaginable problems and phenomena that were much greater than those of radios. The first challenge was that the audio amplification circuit oscillated due to its high amplification.

One example of such oscillation happened at a frequency higher than the ultrasonic wave level, in which case the equivalent circuit of the broadband boost within the circuit was causing it. Another cause of oscillation was an inductive one caused by the configuration of the switch between record and play or errors in the wiring technique. This problem was solved by using a specially ordered rotary switch with a shield plate.

Another difficult problem with the inductive noise caused by factors other than oscillation was the power induction noise of 50 or 60 cycles. This caused the head to pick up magnetic field leakage from the power trance, and also, the wiring from the head to pick up magnetic field leakage.

Other unexpected noise was the metallic clinging sound that the vacuum tube produced mechanically. I solved this problem by suspending the tube with vibration-proof rubber to prevent the sound.

Dealing with such problems turned out to be valuable experience that led to the development of magnetic recorders later on, so considering that encountering these problems at an early stage resulted in the acceleration of tape recorder and video recorder development, and eventually in securing a position unsurpassed by others, I think that we were extremely fortunate.

Taking on Development with Hand-made Measuring Equipment

What type of measurement equipment was used back then in Totsuko for development? Testers were supplied to each staff. Vacuum tube voltmeters could be rented out freely, but I think the company had only one audio frequency oscillator or oscilloscope (electric waveform measurement machine). The oscilloscope was nothing like what you see now — it was just big and tall, and you had to peek into the radar to use it. But we all improvised to measure and create products.

When I was working part-time as a student, I used to handcraft whatever measurement equipment was needed. So I brought them to my office. The audio frequency generator was a Wien-Bridge, an accurate and stable machine. The oscilloscope was small-sized and optimal for waveform observation, hand-made using a 75FB1B-type oscilloscope tube, and I used it a lot from the wire era through to the tape era.

In February, 1949, I had completed assembling and adjusting the wire recorder kit. It was apparently much smaller and of higher quality than the Nichiden (Nippon Denki) wire recorder, so the senior executives ordered me to "stop working on the Nichiden and develop Webster's wire recorder." So, I copied the Webster machine and measured its size, drew up an exploded view and ordered the parts ASAP, and built it up in two weeks.

I was impressed that this machine was designed using quite unique American concepts. By that, I mean that the mechanical section was almost completely made of pressed or punched sheet-metal, and the shafts and bearings were all pressed and swaged. The only screws were the four that held down the motor. It was a complete work of sheet-metal, which, at the time, seemed like a toy to me.

The Webster recorder's wire was not steel but stainless steel (I later found out that it was 18-8 stainless), and its diameter was 0.1mm.

By this time, a massive wave of import goods and information poured into Japan from the United States, and magazines introducing new machine catalogs and measurement equipment were flooding bookstores. Furthermore, interaction between Japanese and Americans became popular, including Ibuka-san and Morita-san, who heard from the occupation forces that there was a machine that produced sound using a tape, who actually went over to the NHK Broadcasting Hall to listen to it. They agreed that the sound quality of the tape was incomparably better than the wire, and decided to start developing the tape right away.

2. Birth of the First Japanese Tape Recorder

Take the Shortest Distance at Full Speed

During the second half of June in 1949, I concentrated all of my efforts on the development of a tape that produced sound and fortunately achieved that goal in less than ten days after I got hold of some ferric oxalate. I don't remember suffering from a sense of failure or frustration during that time. On the contrary, I was enjoying each day, and if one thing didn't work out, I thought, "if this doesn't work, how about this? " and enjoyed my work – feeling the pleasure of solving a puzzle.

Luckily, I was assigned to this development project all alone, so I could work the way I wanted and didn't have to waste my time on meetings and policymaking discussions, which enabled me to take the shortest distance at full speed.

In development, a perfect completion does not have to be the goal. What is most necessary is to reach the goal that you set for as soon as possible and show that your dream can actually be realized.

You must understand that ordinary people can never create something completely in the first try. So you shouldn't fret over small mistakes, and work on reaching the closest possible position to your goal as fast as possible. If you can somehow achieve something, then you can use that as a footing to achieve an even better result. Questions may be resolved instantly once you find a clue.Reaching your goal as quickly as possible and showing that it can be realized is the most important thing.

I believe that my senior executives changed the course of the entire company and decided to manufacture and market tape recorders the moment I told them that the tape produced a sound, which was one of their goals.

As proof of this, they had immediately placed an order with Honshu Paper a long-fibered craft paper containing hemp. There is a saying, "know when to say when," and they actually followed this saying. Once the magnetic powder was produced, we needed a sturdy paper to apply the powder onto, followed by a machine to apply it, and then also engineers to do research on magnetic powder, and also funding for all that. Starting from that day, the company changed its course toward a major transition.

Now that I had succeeded in creating a magnetic tape, I thought that the next step was to create a tape recorder. I analogized from the wire recorder, but it was difficult to pull the tape at a constant speed. Judging based on all the knowledge I had, the driving system could be roughly divided into three systems, - a sprocket drive (film advancing gear), reel drive (wind-up platform) and the friction drive, but since it is not a celluloid film, and the reel winding would not pull at a constant speed, I narrowed my options down to the friction drive. The tape can probably be driven by any method, but I couldn't be sure, and Ibuka-san said he was not sure either because the tape was running through a small gap in the machine he saw.

He must have searched high and low after that, because he came back to me two days later, saying,

"Hey, NHK says they will show you theirs. Go on over and take a look at it."

So I went over to their broadcasting studio to observe their test recording. Unlike the type that Ibuka-san saw, it was a Magnecord portable, designed for broadcasting with a separated mechanical section and the amplifier section.

This machine had just arrived and was set on a 19-inch rack about my height, so I could observe it very closely, which helped a lot. Most of my doubts were cleared here. As I had imagined, the tape was driven by a friction drive, capstan pinch roller method (a method used in standard tape recorders).

The very next day, I started designing the machine section, drew up a diagram and ordered the parts. It required about two weeks for the parts to arrive, so while I was waiting, I drew up two more different machine diagrams that I had ideas for.

The first was a complete prototype intended for nothing more but an operation test, so I could run the tape, but rewinding and fast-forwarding would stop if I didn't help it with my hand, and in addition, the strength of the brake spring used to stop the tape was very difficult to adjust, and gave me a lot of trouble. The motor torque used for driving and rewinding the tape was too weak to rewind the tape reel directly. This "prototype No. 0" was nowhere near presentable to anyone, so I decided to use it in the workshop as tape measurement equipment.

The alternating current induction motor too did not seem applicable to a tape recorder since the type that was used for the telegraph key facsimile had such large humming oscillation that could even be felt in your hand. In addition, the rotating speed would vary depending on the workload, making it unsuitable for a tape recorder that requires a constant speed. So, I asked Ibuka-san whether there was a synchronous motor that rotates silently, and he immediately made an inquiry call to a company called DENON (DEN-ki ONkyo Kabushikigaisha). DENON manufactures professional-use ultra-precision equipment such as disk recorders and cutters for recording and delivers to record companies and NHK, so they provided us with several types of motors that rotated silently and produced a strong torque -- just the type we were looking for — and even promised to support us in the future.

The motor they provided was called a hysteresis synchronous motor, whose rotating rotor is made of one tubular magnet which rotates by the pull of the rotating magnetic field, so it rotates synchronously to the power frequency. The hysteresis motor contributed greatly to our development thereafter, all through the tape recorder era to the video recorder era. of the rotating magnetic field, so it rotates synchronously to the power frequency. The hysteresis motor contributed greatly to our development thereafter, all through the tape recorder era to the video recorder era.

First prototype of the tape recorder

The First Prototype Completed

Now that we had a motor, I immediately changed some of the parts that I had additionally ordered, set the new motor on the parts, and changed the damping brake section and the adjustable section of the winding tension to a felt friction board so that the spring tension can be fine-tuned.

By this improvement, the movement of the machine section became almost satisfactory, so I moved on to adjusting the entire machine including the amplifier. The "first barrier" that I faced was the induction noise from the power (50 cycles). When the machine was all spread out on the desk, the motor and the power transformer were set apart from the head, so there was relatively small induction, but now that they were all packed into one box, the sound sources got close together, and induction increased inversely by the square of the distance. This was the first time I set the parts up in the form of a product, so I learned the hard way that unexpected things can happen at unexpected moments.

The electromagnetic induction that the head was picking up was resolved by covering the head with an electromagnetic shield and wrapping the sound source motor and power transformer with a copper band for the alternating magnetic field that was radiating externally to cause a one-turn short-circuit (short-circuit caused by a single wound coil) to minimize the hum as much as possible.

Know-how Taught by Ibuka-san

Tape recorders reproduce sound by drawing a signal line from the head to the high-level first-stage vacuum tube for amplification.

The "second barrier" was the wiring of this signal line. The problem was that the line picks up magnetic induction from the power. Even if the head section was short-circuited, the humming noise was the same and would not go away, so the wiring was picking up the hum noise. My colleague, Katsuzo Otake was helping me with the wiring at the time, so we searched around for the spot where the wiring was picking up noise. Otake-kun was in charge of wiring the radio, so he tried wiring with a thin Litz wire and then covering the wiring with the metallic mesh used on the exterior of the shielded wire, struggling to remove the noise.

Ibuka-san happened to drop in and asked us, "what are you doing?

" We told him, "we can't get rid of the noise." Then he instantly said, "oh, that. Try twisting the lead wire like a spill."

As mentioned in SONY History book "Genryu," Ibuka-san worked on developing submarine detectors during the war. So he knew from his experience that "noise from the wiring can be prevented by twisting the lines." This solved most of the noise problems.

"Okay, Let's Make the Product"

This is how we completed prototype No.1 in September 1949.By that time, tapes and small-sized tape recorders were coming in from the United States, so we had more chances to see the actual products. I measured the tape and obtained data as well. The first tape I saw was gray paper covered with a black magnetic substance, which was surprisingly moist and smooth.

We also had about 10 new graduates employed for tape and magnetic substance research, and my workplace became a hive of activity.

Ibuka-san and others on the sideline requested recording devices that could record longer with better quality, so I had trouble developing the concept for the next machine. Around this time, Ibuka-san often brought visitors to my workshop and asked me to demonstrate the machine. In particular, when Mr.Katsuma Tani (founder and first president of TEAC Corporation) visited, he saw the plan for a one-hour reel that I was working on and said, "did you design this reel? You have a great sense of design."

From his words, I determined that a tape recorder capable of recording a full hour would be my next development project. I wanted to make a product, not just the reel, which would make him say that the tape recorder itself was made with a great sense of design as well.

About three months later, in January 1950, the prototype was completed. It was a completely original tape recorder by Totsuko, a breakthrough from just mimicking American products in the aspect of both technologies and in performance.

In the meanwhile, preparations for selling the product were proceeding steadily. Totsuko and Nippon Denki obtained a patent for the alternate current bias method of magnetic recorders from Anritsu Denki. In addition, Totsuko registered the name "Tapecorder" as its product name. At the same time, tapes made by Totsuko were to be called the "Soni-Tape." "Soni" stood for "sonic," in other words, "sound."

The Soni-Tape, which was almost like an indication of the future SONY, was about to make its debut as early as 1950. Preparations were quietly but steadily proceeding.

When everything was ready for production and distribution, the long-awaited Tape recorder G-type was completed. By the way, G stands for "Government." My design plan was no longer a prototype plan. It had been rewritten as a mass-production plan and was handed over to the production team. The production manager at the time was Mr.Kazuo Iwama, who later became the president of Sony.

The production system for the tape recorder's main unit was being prepared, and was to be headed by Mr.Masao Kurahashi who had come from a company called Yakumo Sangyo, but the production system of the tape was not ready yet, so in November 1949, Mr.Keizaburo Tozawa (later president of Sony-Energytec Inc.) was summoned as the tape development and production leader.

This was how the tape team was established, and with reinforced personnel, development outcomes improved dramatically. At first, some suggested importing the tapes from the United States to sell with the tape recorder, but Tozawa-san and the team worked very hard to complete the Soni-Tape on time whatever it took.

Japan's first G-type tape recorder

We produced not only the general-purpose G-type, but also professional types, and those were delivered mainly to broadcasting stations.

After Shiro Hojo entered the company, types BS-1 and BS-2 floor-standing type broadcasting tape recorders were developed and went on sale at around August 1950.

At around the same time, Katsu Inaga developed portable tape recorders KP-1 and KP-2 for broadcast use.

Japan's First "G-type Tapecorder" Goes on Sale

In 1950, the entire company was packed with the excitement of the upcoming release of the long-awaited tape recorder.

Research on physicalities such as the magnetic substance was undertaken in cooperation between many companies and schools including Tohoku University, Tokyo Institute of Technology and Tohoku Kinzoku Kogyo, the predecessor of TOKIN Corporation), and promotion was actively conducted in full scale targeting the broadcasting sector and public offices, newspapers, and magazines. I too was asked to demonstrate the machine in trade shows and presentations.

In the March 1950 issue of "The Mainichi graphic" issued from the Mainichi Newspapers printed articles on the development of Tape recorder A-type, G-type and Soni-Tapes.

This was the first promotion that Totsuko ran, and the news having been released before the product went on sale, it created much buzz. Totsuko must have been a company that was ahead of its time even back then.

In May, we presented the Tapecorder G-type to Emperor Showa.

While the entire company was working progressively on the release of the product, the critical path was the production of the tape. Tozawa-san recalled that period:

"When I took over the tape development, I felt that this was a totally new project for me. No one in Japan had produced a recording tape yet. If nobody had done it before, then we had no disadvantage. This was a really interesting project, so we had to achieve it somehow. They said the States started producing their tapes after the war, so we were only about three years behind. There was no reason we couldn't make a better product.

Developing the tape recorder G-type

As for the recorder, we had imported samples, and prototyping was underway at a good pace using American tapes. But as for the tape, we couldn't figure out how it was made just by looking at American samples, and moreover, we had no idea what kind of equipment was used to make it, so we had to create everything from scratch. This was going to be a tough task, but now that I had assumed responsibility, I had to produce the tape by the time the recorder went on sale."

That was the mixed feelings he had when he started developing the tape. In 1950, many new recruits joined the company to work on the tape development. Takuji Tachikawa, Akio Amaya, and Shinichi Tokumoto were assigned to the tape team.

In less than 10 months after the team was launched, we were able to create a tape equivalent in quality to the high-quality American tapes, and finally, Japan's first tape recorder GT-3 and the first type10 reel Soni-TapeKA was released in August 1950.

Commemorative Recording on the First Anniversary of "Tapecorder" Research

Exactly one year after top executives Ibuka-san and Morita-san decided to develop the tape recorder, we got together at Morita-san's home in Setagaya to record our voices in commemoration. In 1995, we found that tape 45 years later in the warehouse, and when we replayed it on the G-type, it replayed a very clear sound.

The paper tape and the G-type machine that we created retained the sound after half a century, and the machine worked without any malfunction. It was a very moving moment.

I will note a part of that recording here.

Ibuka-san: "August 6th, 1950. This is the first anniversary since the Tapecorder project started, and there are 14 of us together here today to recall the past year, and imagining the next year, discussing many topics to work even harder. I am looking forward to comparing the quality of this recording today and the recording a year later. I have high expectations."

Morita-san: "This is a G-type, 15-inch speed machine. However, this is the second G-type that was adjusted by Kihara-kun. Since last year, when he pasted the OP with a brush and produced sound, we have succeeded in creating this machine in just one year, so we are recording our voices in commemoration. We would like to record our voices every year as data to make this our history. This was Morita speaking."

Iwama-san: "Uh, I am Iwama. Uh, the Tapecorder has reached a very tough level, so we have to work together even more than before, and toil to develop things much further."

Me: "Uh, this is Kihara. Looking back at the year before, I am impressed at how we started off with such low-quality stuff and managed to achieve this moment. I played around with both the powder and the machine quite a lot, but I couldn't make anything handsome, so I am a bit disappointed about that. Well, I am looking forward to the next year to see what we will have produced by then, but I believe that we all have to work much harder, or else, production of anything better than this would take a very long time. However, I would like to be able to happily say a year later that we have made a very good product."

Witnessing the fact that the machine can still replay these voices clearly half a century later, I am deeply moved to be able to reaffirm that there is every reason the magnetic recording technology is now being widely used as video and computer memory.

CHAPTER TWO

SOON HOT, SOON COLD

The Entrepreneurship that Masaru Ibuka, my Mentor, Taught me

LEARNING "MACHINES"

Competition was quite steep for entrance exams to get into science colleges in 1944, with only one in 40 applicants successful. The reason for this was because science students were supposed to be exempt from military service so that they could work on the development and production of military supplies. It was natural that a lot of people decided to go for the science colleges simply because they didn't want to be drafted to go to war. Regardless of whether they had studied science in high-school or not, most students applied to science colleges and took the entrance exams. I learned about this a decade or so after the war was over at a school reunion I attended. I browsed the alumni list and was surprised to see that only six of us had taken up technology-related jobs, and the rest all joined white-collar companies such as banks and trading companies.

Among the few tech alumni was my friend Yukio Hasegawa. He attended the a Faculty of Science and Engineering at Waseda University and then became a professor at the Institute of Systems Science Research, and now, he is active as a big name in the field of industrial robots. Since SONY also produces small-sized robots for industrial machines, we have maintained a long-standing relationship, and he has helped me a lot over the years.

Despite the severe competition at the time, I was fortunate enough to manage to enroll in the mechanical engineering department of Waseda University. I liked playing around with machines to begin with, so I remember enjoying doing the lab-work and studying drafting. In particular, I learned how to make pots and pans from sand-molds in the casting lab, and that was when I first learned that you can use beer to solidify sand. In fact, one of my relatives ran a casting company, and I used to visit the workshop and play around with the machines. But this company handled lightweight metals, for which the process of mold-making was totally different. So, I was very impressed to learn that there was a different method for iron casting.

I also learned about structural dynamics in the mechanical engineering school, working mostly on the calculations of heavy machinery for bridge designs, or oscillation analysis of 4-cylinder and 6-cylinder engines. These use a lot of mathematical calculations that involve calculus, so it was a very fun and interesting class for me. I also never missed my math class. At the end of class, the professor would always write a problem on the blackboard and ask, "anyone want to solve this? "For example, if he wrote an integral equation, I would always take up the challenge and enjoy walking up to the blackboard and writing the answers.

In front of the Okuma Amphitheater at Waseda University

As for electrical engineering, the class mostly consisted of learning how to handle DC/ AC motors related to machinery, and lab-work on how to use them. Needless to say, there was hardly ever a class on electronics at the time, which is the mainstream now.

STUDENT MOBILIZATION

The year 1944, when I entered University, was also the year prior to the end of the Second World War. It was a very urgent period in which students had no time to be taking classes at the university and studying leisurely, as we were mobilized as working students being sent to various places. I once was sent to a truck repair factory to disassemble malfunctioned engines, replace piston rings and the valve lapping, and adjust them until the trucks could run again. The factory manager and the team leader would adjust the distributor ignition timing and complete the task in the end, but to me, repairing automobile engines proved to be an important experience.

I also spent a few days helping with the road expansion work in which wooden houses were taken down to prevent fire from spreading in cases of aerial attacks. This was hard labor, and we had to remove all of the roof tiles, remove the walls, and once the house was stripped down to just the framework, you had to hook a steel rope on the top of the roof and pull it down with everybody else. I remember doing this work thinking it was all for the good of the country.

And finally, I was sent to a factory called Aichi Tokei Denki, located in Atsuta, Aichi prefecture. The factory produced engine parts for airplanes. My job was to carve out cam shaft material for engines using a 1.8meter lathe, with a daily quota of at least five pieces. This lathe was a large-sized machine tool about three meters in width.

I would attach a 1.5 meter wide, 8 centimeter-in-diameter round steel bar to the chuck of this lathe, and using a turning tool, cut the bar in one shot all the way to the thickness of the shaft, leaving room for the cam. Since the bar is unannealed, the steel would be shaven off the surface in thin plates like wood shavings, which would be oxidized by the heat generated when being carved off, and shine in a bright ultramarine color. I always admired that view.

In small downtown factories, machine tools were mostly small too, so it was the first time for me to operate such a powerful tool by myself, and it really was a valuable experience. I was taught how to sharpen the turning tool so that I could do it myself, and I also learned tips on how to attach the turning tool to the lathe, and was able to carve out steel in a consistent manner.

I also mastered how to use the center drill, which is used to set the center of the steel bar. After a day's work, pitch-dark shavings piled up in heaps. These were cleaned up by the next morning, and I was told that they were taken to the furnace to be recycled.

This task continued for about a month, after which I was assigned to help with the processing of the milling machine (a machine tool frequently used along with the lathe). This too was a gigantic piece of machinery, whose diameter was an amazing four meters, and the drill section about three meters in height. The entire machine was set in a temperature-controlled room, where I worked in. The work pieces were something that resembled a gear box about a meter high and two meters wide. There was a dial face surrounding the turn table, and my job was to read aloud the numbers of the dial, and turn the fine-control dial to adjust the primary dial. This milling machine was made in Switzerland.

I was never told what was being made from these processes, but during the nights at the dorm, we students exchanged information that each of us gathered and based on that knowledge, we assumed that we were building a prototype German liquid-cooling engine, and that the blueprint for that engine was probably brought from Germany by submarine.

I wonder if the liquid-cooled fighter plane ever took off. I wonder if what we did was of any good back then. There were constant air-raids and bombings on the factories, and considering that it was a year before the war ended, I doubt it. I can't help but think that the war was so meaningless and only ended up taking everything away from so many people.

THE FIRST AIR-RAID

In October, the course of war seemed to have taken a turn for the worse when enemy aircrafts started attacking the industrial areas of Nagoya and Atsuta almost every day. In the evening from our dorm, we could see red pillars of fire shooting up here and there in the direction of Nagoya. Since we were young, we almost enjoyed the view of the fire far away for the first few days, but as the days passed, we realized that the bombings were coming gradually closer to where we were.

I heard later that Nagoya Castle had been completely burned down, keep tower and all, because of the air-raids at around this time.

I had actually climbed up the castle just a few weeks before that, when there wasn't much work to do in the factory. The castle had maintained the same magnificence it had since the Tokugawa Era, and the golden "kinshachi" dolphins reached toward the sky. I was lucky to have seen the real castle then. It really is unfortunate that a historical monument was lost overnight.

A friend obtained information that a full-scale bombing would soon be aimed at Atsuta, and that very night, a bomb was dropped right by our dorm. We scrambled into the bomb shelter and waited until things quieted down, then went to the Aichi Tokei factory. The factory in which I worked in was intact, but the factory next door had caught fire.

The students all got together and agreed that, since the teachers were not there, we should escape as soon as possible, and decided to take the train the next morning to return to Tokyo. There was no time to contact the factory manager, the teachers had returned before us, and we just thought if we were to be scolded later, then that's that. But more importantly, we were worried about whether the railway was still operating. Trains could run on just coal and water as long as the railroad was not damaged. So, though the train was overcrowded, we managed to return to Tokyo safely.

Once back in Tokyo, I was relieved to see the serene and safe state of the city. We got to see our teacher and when we consulted him, we were told that we should continue production for the war, so a few days later, we went back to Nagoya. But everything had completely reduced to burned-out ruins.

The factory where I worked had been completely wiped out by the bombing, and I couldn't even find that gigantic lathe, though I searched for it. The building where the lathe was set up, which was a little away from the main building, was still there, but the machinery inside was all burned down, and the expensive precision machinery had been turned into ferrous scrap, which made me very sad. The Tonankai earthquake that had occurred on December 7th had forced the buildings to collapse either fully or partially, and since it had been decided that it was impossible to continue production, I was sent back to Tokyo to wait until my next assignment was determined.

But into the next year, no news arrived at the university, and in the end, the student mobilization was terminated due to the deterioration of the war.

THE GREAT TOKYO AIR-RAIDS

After New Year's Day of 1945, we started to hear a lot more air-raid sirens. My house was located in Higashi-Nakano, which was convenient to commute to school which was in Waseda, but since it was relatively close to the center of Tokyo, it was a dangerous area which was targeted in air-raids. In March, enemy planes attacked and strafed the Shinjuku area, so since then, the community executed firefighting trainings and bucket-brigade trainings every day.

Finally, on the night of March 10th, a formation of The Boeing B-29 Superfortress bombers attacked, and the center of Tokyo was burned down by incendiary bombs. Before and after this attack, formations of enemy planes passed high above Tokyo, probably close to the stratosphere, even in broad daylight. I could see a dozen silver planes in the middle of the clear blue sky. They must have been taking pictures of Tokyo to use as data for the next attack. I could also see clearly that a few Japanese planes were attacking the enemy planes to prevent any further bombings.

At the same time, anti-aircraft artillery must have been used, because I could see small flashes of lights near the enemy planes, and a round ripple that spread slowly around the light into the blue sky. It looked very beautiful to me, as if I was looking at a mystical painting. When the bombshell exploded, the air propagated from the center as a compression wave, and the refraction of the light must have made it look like the contrasting rings spreading out into the sky.

The pilots were fighting for their lives, and though I did think it was imprudent, I couldn't help but watch the scene in amazement that such awe-inspiring beauty existed in war.

I was observing the situation from the top of the roof as a lookout, but the next day, I saw a Japanese fighter plane fall from the sky dragging white smoke behind it. I cursed at the sight, then looked up and saw something that looked like a white stick falling straight down after the plane. It grew bigger and bigger as it neared the ground, and I was shocked to see a human being on the tip of it. The parachute had failed to open, and the pilot was falling straight to the ground. What's more, the pilot was saluting with his right hand. It definitely looked like that to me. He was facing the direction of the Imperial Palace. I thought I almost heard him say, "Long live the Emperor! ''

As predicted, the newspapers in the following day reported the tragic death of the pilot. I remember that despite such a chaotic and severe situation where everything was lost, the papers were strangely delivered almost without fail every day. The press must have been working around the clock to write the articles and find valuable paper to print them on to deliver them to homes. I was impressed by their commitment to their work.

ANOTHER AIR-RAID

On the night of the 20th, there was another attack by a huge formation of B-29 planes, and my house in Higashi-Nakano was burned down by the incendiary bomb. I was up on the roof as soon as the air-raid siren sounded, and was looking up at the night sky when I heard the signature humming sound of the B-29 formation. This sound had been constantly broadcasted on the radio for some time, and having memorized it as a distinguishing sound of enemy planes, I knew right away.

Until then, enemy planes would only fly either high up in the sky or far away, so I had never heard them myself, but now, they were actually pressing in on me. Their bodies, lit up by search lights, looked less silver and more like giant transparent objects in the night sky.

I was likely shouting "enemy planes, enemy planes" from the top of the roof, but the incendiary bombs must have been dropped by then, because in a few seconds, immediately after the planes passed above my head, balls of fire rained down on me as if a huge firework had exploded opened above me.

I frantically tried to get down from the roof, when one of the bombs dropped right into the middle of my house with a "pop," breaking through the roof tiles. I don't remember hearing a huge sound. It was very close to me, only about five meters away.

Thinking that I needed to extinguish the fire, I grabbed a bucket of water and rushed into the house, but the bomb was half buried into the tatami and breathing fire. It was so hot I couldn't get near it even to pour water onto it. I gave up on extinguishing the fire, and carried as many valuables and movables as I could fit into the bomb shelter, covered them with soil, and temporarily evacuated to the temple that was behind my house. The incendiary bombs were normally dropped only about 20 meters apart, so extinguishing one or two of them wouldn't help because the whole area would be on fire. I was rather grateful that the bomb didn't drop directly on me. It really was a small mercy on my behalf.

I learned later on that one of my classmates from junior high-school was hit directly by an incendiary bomb, and another was killed by a cracked piece of the anti-aircraft artillery bullet.

PRODUCING AND SELLING ALUMINUM PANS

Since my house was burned down, I had to find a safe place to stay. An aunt of mine who lived in Ichikawa, Chiba Prefecture, had always told us to come over whenever the situation turned worse, so my parents and I decided to go there.

When the war began, I never felt a shortage of goods. I remember my parents buying me a brand-new uniform and cap when I entered Waseda University, and that we all were happy about it. We even took a commemorative photo. Though they were being rationed, I never felt a lack for commodities such as socks and shirts. Even shiny new leather shoes could be found if you looked for them.

But by the time the air-raids and bombings intensified, everything had become extremely hard to come by, including food. It must have been around this time that people started bartering goods instead of money, like trading clothes for rice. A relative of mine who lived in a village near the coastline of Kashimaura, Ibaraki, worked as a village office worker as well as a farmer, so he didn't have much trouble with food compared to those who lived in the city. He told us to come over and pick up his potatoes, so I would ride my bicycle to procure food each month. The best feast we could hope for at the time was dried potatoes and potato candies, which I stacked up high on the back of my bike.

It was clear that the war was coming to an end. I wanted to know what was going to happen to the school, so I asked around, but I couldn't find anyone at school, not in the office nor in the teacher's room, and I couldn't find any friends either, so I decided to give up on going to school for a while.

I was hanging around for about two months, and since I had nothing to do, I went to the casting plant that my aunt's family owned to help there. At the time, we were mass-producing pots and pans out of aluminum mold which were sand-finished before putting them out for sale.

The war was still going on, but orders from the government had stopped, so the plant discussed what to make next, and decided that pots and pans would be a good choice since everyone had lost their cooking utensils.

True, rice and vegetables could be obtained with some effort, but without pots and pans, you can't cook them. There was no replacement for these things, so our quick-made pans sold very well.

Back then, commonly used cooking tools were mainly clay charcoal stoves with pea coal or coal briquettes. Since it smelled a bit of pit coal, we would take it outside the kitchen or to the back yard and use it outside. Baths were heated by coal in a coal-fired kiln, so my family rarely had to go to the public bathhouses. I think coal was delivered whenever we asked for it.

Charcoal was essential to daily life, and it was a normal sight to see people boiling water on a charcoal brazier. In summer, an iceman would come to sell ice with a pillar of ice on his cart and sawed off blocks of ice with a huge rough-edged ice saw.

Icemen can be seen occasionally in town even now, half a century after the war. Since we had a wooden refrigerator back then, the iceman would come every day to deliver ice. On a hot summer day, I remember my father making home-made ice cream, which was very delicious. Even during the war, we were able to live a decent life using a bit of ingenuity, so I guess the shortage of goods wasn't that serious around this time.

There was gas, but during the war, it was hardly available for use.

Thankfully, during our temporary stay in Ichikawa, we didn't feel much inconvenience. But the school had been closed down, and there were rumors of battles on the mainland. Ultimately, I was left to await the arrival of that red draft notice.

THE ATOMIC BOMB AND THE END OF THE WAR

One hot summer day, an emergency radio broadcast started with a tone different from the usual one. At first, I couldn't understand what was being said. It said something like "a new type of bomb seems to have been dropped in a town in the Western region. We are still investigating, so please do not panic or overreact until the next update."

The news was censored around this time, so there were news that made lost battles seem as if they were victories for us, and news with ambiguous expressions, so I was surprise to hear the word "new type of bomb." Everyone in town was talking about this for a few days, and gradually, it became clear that the scale and force of this bomb was unimaginably large. People became afraid that Tokyo might be the next target, and many people left town to evacuate to the countryside where it was safe. My family prepared for the inevitable and decided to wait for the war to end.

When I heard that the second atomic bomb was dropped in Nagasaki, I wondered what the point was of continuing war at such great costs, and I wondered why the military was calling for honorable defeat. As I had anticipated, the Japanese government chose unconditional surrender, and the war ended.

The day before the end of the war, the radio decreed that "there will be an important word from his Imperial Majesty to the people of Japan, so everyone must listen." I doubted that it would be about the mainland battle, and spent the day feeling uneasy until August 15th came. I had mixed feelings when I finally heard the broadcast.

I never went to the frontlines and never experienced the suffering of real war, but people who lost their families in the war, people who were injured in the war, people who suffered severely in the war must have been angered beyond imagination.

But in reality, there was no time to be thinking about anything. How are we going to live out the next day? What kind of life was waiting ahead? How was the world going to change? Unsettling days were about to start.

circa 1929, in front of my house
in the Shanghai Japanese Settlement

The rumors then said that first, the American soldiers would make a landing en masse. Then, they would probably pillage the city. So we must hide our women and children. Everyone was panicking.

Next came the food shortage. During the war, we all shared what little stock we had evenly. But once the war was over, the "for the country, for victory" mindset disappeared, and everybody started to stock up just for themselves, and the black market was born. Japan fell into a state of anarchy where black markets and black upstarts flourished.

The occupation of Japan by the United States passed without much chaos, and knowing there was no need to worry about being pillaged, life during the occupation gradually stabilized. The currency turbulence was settled by the printing of the new yen, but the black market persisted.

LOSING MY FATHER

When my father was young, he wanted to become an engineer, and had intended to go to an industrial school. But he suffered from lung trouble, and couldn't study very hard, so he gave up becoming an engineer and followed the commercial course at Waseda University. After graduating from Waseda, he joined Nippon Yusen (NYK), and apparently, he primarily handled the allocation of ships to Southern sea channels and the Pacific Ocean channels, as well as cargo operation plans.

When I was in junior high-school, he worked in the Nippon Yusen Headquarters next to the Marunouchi Building, and later during the war, he was the branch manager at the Yokohama branch.

I think that Nippon Yusen was then a company to be admired, like Japan Airlines (JAL) or All Nippon Airways (ANA) is now. You could travel overseas on a luxury cruise ship, and you could even work overseas. I too, travelled by sea three times between the age of three to seven. I had accompanied my father who was assigned to

Shanghai, and I remember the ambience of luxury cruise ships such as the Chichibu-Maru, Nitta-Maru, and Tatsuta-Maru. There was a swimming pool on the ship, and there were many people on the tennis courts and playing table tennis. I would run around the deck alone, and once, after going back and forth in the lower decks, I got lost and had a shipman take me back to the deck.

The most vivid memory I have of back then is how lavish the dinners were and how good they tasted and smelled. On cruise ships, the best cooks selected from all over the world would use plenty of expensive ingredients, so the food must have been great indeed. Their soups were particularly exquisite, and I can still recall the smell vividly. Sometimes when I am travelling in Europe, I catch the smell of that soup, and it reminds me of my cruise.

In the good old times of the early Showa era, my father must have been leading a much more luxurious life than the average business-man. I don't know how he was at work, but at home, he was a good father, and he would buy anything I asked forEven when I didn't ask, he would find the latest imported toys like model railroads and build-it-yourself toys and would happily open the box for me and tell me how to play with them.

Since he wanted to become an engineer, he must have liked to touch and play with those elaborate mechanical toys, too.

My father was much more skillful than the average person. Having lived in China, he became very interested in art, enjoyed calligraphy and painting himself. He purchased a book by Wang Xizhi, the Sage of Calligraphy and practiced calligraphyand eventually becoming qualified as a master calligrapher, choosing Shunsui as his pen-name. In addition to the cursive style, he also mastered the demotic semi-squared style and seal-engraving style, and would engrave in reliefs the seal-engraving style characters none of us could read on Chinese wooden plates and frames.

As for his other artistic hobbies, he would take photographs, develop them and print them all by himself. He created a wooden window frame in a small room in the house so that he could cover up all the windows and doors, installed a water supply and attached a light controlling device to the desk lamp to create a perfect darkroom. When I was a child, I would ask my father to let me in when he was printing pictures, and would stare amazed at the images gradually appearing onto the dry glass plates or enlarged printing paper.

My father used to send his pictures to the Asahi Shimbun's Japan Photo Annual, in which his photos would almost always be selected to be printed. He took pictures of the scenery and people in China, as well as news photographs of the Shanghai Incident which took place in around 1932.

My father was very talented at sports, and whether it was iron bars or anything else, he would exhibit techniques that would have embarrassed professional athletes. He would show me a back hip circle and then a giant swing at the iron bar in school, and one day, he created a sand pit and an iron bar in our back yard to teach me how to do them. I had to practice the back hip circle every day from then on, and I practiced so hard that I got many blisters on my hands, and finally managed the back hip circle, but I gave up on the next target, the giant swing.

Back then, there were a couple dozen cups and plaques for tennis and table tennis competitions, which were apparently trophies won by my father at company competitions. Those were all taken away during the war since all rare metals had to be contributed. After that, the war deteriorated, and our house being burned down by the Tokyo Air-Raid, we were left penniless. It was something that had been anticipated, but still, it must have been hard on my father to see his life's work vanish like a bubble. He frequently suffered from a pain in his stomach, and he started to take stomach medicines regularly.

Near my home in Higashi-Kanagawa, in my father's hands

After the war ended, he had to go out frequently with the occupation army officers on his business with the Yusen, and due to excessive drinking and smoking, he vomited a lot of blood one day, and died quietly, as if vanishing into thin air.

MY HOBBY SAVED ME

Now came the one of the biggest turning points in my life. My father left usretirement benefits, but unfortunately, this coincided with the conversion to the new yen, and all of the cash was rendered essentially worthless.

In other words, I had to work in order to get a hold of the new yen. My options were to either become a laborer, or work part-time as a student. I couldn't resign from the university that I worked so hard to enter, so my only option left was to work part-time as a student. But at this time, part-time jobs for students were either jumping on pickup trucks and helping people move, or cleaning buildings, so I quit in about a month.

One day, a friend at school told me there was someone looking for a radio, so I made him one. That became the trigger, and people who heard about me started asking me to make radios for them. I had made trans and motors on my own since junior high to use in my model railroad, and had been making high-frequency first stage radios without much trouble just by reading magazines such as KODOMO NO KAGAKU (Children's Science). After the war, with nothing to do since school was practically closed down, I was making short-wave radio receivers and electric phonograph pickups on my own. So, though an amateur, I could make a 5-tube super heterodyne radio without looking at the circuit diagram.

I started to frequently visit Kanda, an area that is now called Akihabara, to purchase parts. It being right after the war, the place was little more than a few street vendors standing amidst the ruins, and their electrical parts were mostly junk or recycled parts, but the strange thing was, there were occupation army surplus parts piled up in heaps, too. I was grateful to find American vacuum tubes there.

In order to make radios and electric phonographs, you need a wooden box and a turn table, a pickup and a dial display. What makes the product look especially good is the dial display. What I wanted then was displayed at one store. It was a type of a traverse dial onto which light shone from the side of the glass to make the display light up, and it was very, very beautiful. Ever since finding this, all of my radios and phonographs used this beautiful dial.

MEETING IBUKA-SAN

I worked part-time for about two years to help my family, and finally managed to graduate from Waseda. Next, I had to worry about work. From the various fliers I found at school, I chose Nihon Denki, Nihon Musen (Japan Radio Corporation) and Toshiba and started job hunting. Looking back now, these were all electric companies. I didn't know any of the names of mechanic companies, so I must have unconsciously chosen electronics companies whose names I had seen in Kanda frequently.

One day, a recruitment flyer pinned on the information board at school caught my attention. The information, which said "Recruiting students. Masaru IbukaTokyo Tsushin Kogyo co. ,Ltd (Tokyo Telecommunications Engineering Corporation)" caught my attention because I saw the name "Masaru Ibuka."

Ibuka-san held lectures at Waseda University every once in a while as a part-time lecturer. Despite being a mechanical major, I liked electronics more, so whenever there was a lecture by someone in the electronics field, I would go and listen. Unlike other lecturers, Ibuka-san's lectures were very practical, very easy to understand and interesting.

I think I heard his lectures maybe three times. I had the impression that he was working in his company while at the same time researching new things.

Since I made radios and electric phonographs as a part-time job, I knew about the name Tokyo Tsushin Kogyo. I had used a lot of pickups, turntables and dials made by them.

Knowing that Ibuka-san was someone from the company that made my beautiful dial display, I became more interested in the Tokyo Tsushin Kogyo. I made up my mind, "I must visit Tokyo Tsushin Kogyo and see what kind of a company it is." I consulted the school right away, prepared my resume and application form, as well as a reference letter, and decided to visit the company a few days later.

There were many other recruitment flyers on the information board, but I don't remember seeing any one of them. That shows how destined it was that I saw Ibuka-san's message.

I walked from Shinagawa station to the company in Gotenyama to take the entrance exam, but no matter how much I looked, I couldn't find the company. I finally asked a passer-by, who told me, "it's behind that house there." The company was less than a tenth the size of what I had expected – a tiny, tiny wooden shed.

The only factories I knew of were Aichi Tokei Denki, Nihon Musen, Toshiba and my aunt's Kihara Casting Factory, so it was no wonder I had so much trouble finding Tokyo Tsushin Kogyo, which was less than half the size of Kihara Casting Factory, which I had until then assumed was the smallest factory ever.

I found the entrance of the Tokyo Tsushin Kogyo, which was at the back of the house the passer-by directed me to, and was ready to take the job interview. The interviewer was Mr.Akira Higuchi, who later became the Vice president of Sony. He asked me many questions while looking at my resume, but the first thing he asked was, "You're supposed to be a machinery major, but what you've written here is all about electronics, like making radios and electric phonographs. That's interesting."

I think I said something like, "I like playing around with electronic machinery," but I regretted that I wrote my resume badly, so I was very relieved when I received the acceptance letter a few days later. I was told to "drop by if you have time before starting to work regularly at the company," so when I didn't have school, I visited the company frequently, and had the chance to meet Ibuka-san, too.

From these visits, I learned what kind of work the company did, as well as their research policy, and started to think that this company might let me do something interesting. So I thought, "Alright, rather than joining a larger company, I'll work here and see how it turns out." That is how I decided to join Tokyo Tsushin Kogyo.

THE UNFORGETTABLE WORDS OF IBUKA-SAN

Ibuka-san is talked of as being a great person in many aspects, including as a creator, business manager, educator and visionary. So I am not going to review that here. I will only introduce some episodes about his personality that deeply moved me, and about his spirit that shook my soul as an engineer.

This episode took place much later, when I was around 27 or 28. I had a problematic habit of developing a fever after working too hard, so I frequently had to take days off of work because of it. Back then, Totsuko didn't have clinics, much less welfare facilities for employees. So whenever I had a cold, I would go see a doctor in town and get a prescription.

Ibuka-san must have seen me as a fragile young man. He told me,

"I will refer you to a doctor. Take this reference letter and get yourself checked out."

And handed me an envelope.

I immediately went to see that doctor, and spent half a day having my body checked all over, even having x-rays taken. When I was leaving, the doctor handed me an envelope addressed to Ibuka-san. It was sealed, so I didn't know what was written inside, but after reading it, Ibuka-san said, "Good. I was worried, but there's nothing wrong with your lungs. I'm relieved."

I didn't understand why Ibuka-san was so relieved at the time, but I could see that he was really worried about my health. I also suspect that he must have paid for the fee of my checkup.

Another episode is when I was a manager making tape recorders. Since my job is to make things, I get too concentrated on making the products as soon as possible, and nothing much else comes into my view. My desk is always in chaos.

Ibuka-san dropped in to check on me laboring away single-mindedly in the work clothes provided by the company.

"This is what I made," I showed Ibuka-san the prototype machine, when the heated soldering iron that was placed offhandedly touched my work clothes. But I continued explaining unaware. In the meanwhile, the work clothes that were touching the soldering iron started to give out burning odor. My boss who happened to be there warned me, "you shouldn't leave the soldering iron out like that." By that, he meant that the soldering tip shouldn't be exposed." Try to tidy up a little more."

He has very reasonable, since it was also dangerous to leave things out like that. But Ibuka-san said, as if to stop my boss from scolding me,

"Oh, don't worry about burning your clothes. I will provide you with as much clothes as you need."

And he also said, "rather than that, concentrate on making products much faster." I was surprised, and at the same time, keenly felt "wow, he really is on a different level." These are some of the words fromIbuka-san that I can never forget.

This is another episode that happened when I was a manager. I used to commute to work in my own car and, eager to start work as quickly as possible, I would often park my car on the street behind the company. Ibuka-san would drop by two or three times every week to check up on me, and when I was explaining to him about a new machine I made, I heard an announcement saying, "License plate Shinagawa, number XX, please move your car."

"Oops, that's my car. I've parked illegally, so I'll go and move it." I told him, and while Ibuka-san was waiting for me to come back, I moved the car. Then I went back and resumed my explanation, but a few days later, General Administrations told me, "there's a parking space for executive officers in the basement. One of them has been assigned to you, so please park your car there starting tomorrow." I was assigned an executive officer's parking space! Ibuka-san never told me that he gave me a parking space. But he must have meant "don't worry about your car, just concentrate on your job."

Here is another episode. One day, Ibuka-san was watching me draw up a plan.

"How long will it take for this plan to become a product after you order it? "

"If I rush it, one week. Ten days on average."

After this conversation, Ibuka-san must have assumed that it will take about ten days after a plan is drawn up until it becomes a product. Again, a few days later, engineering works department staff came over and set up a workshop in the corner of the floor on which my office was located.

"Why? "

"Ibuka-san told us to."

So, he made the workshop just for me. That wasn't all. He hired five craftsmen who handled lathes and milling machines. That meant that I no longer had to draw a plan and make a blueprint. All I had to do was order the men to "make this, with such and such dimensions." I could make products freehand.

Ibuka-san really wanted to see the product work. Me too, I wanted to build it up quickly and make adjustments. He never said that he "made the workshop so that you can make the products faster." But I don't think it's an exaggeration to say that this workshop enabled me to make products faster than our competitors.

Ibuka-san must have been ever eager to see the deliverables as soon as possible, while at the same time, thinking about the priorities of what should be done at this particular moment.

RECRUIT NUMBER ONE

When I joined the company in April 1947, my direct superior was Mr.Kunisuke Hamada. My first job was to determine the resistance value required for the sensitivity setting of the tube voltmeter, to search the resistance closest to that value, and to solder the parts and build them up. This tube voltmeter for VHF AC measurement was completed by Mr.Junichi Yasuda when Totsuko was still Nihon Measurement Company and was said to be one of the few measuring instruments that Japan could boast about having brought the world. The feature of this measurement probe was that it had a very high internal resistance, which eliminated most of the disturbance on the measured object. For the measurement circuit, it used a balanced bridge method which made it highly sensitive, and for its accuracy, we used it a lot on our development and prototyping. At the time, this tube voltmeter (we used to call it "val-vol") was mostly delivered to the Ministry of Communications (now the Ministry of Posts and Communications).

Totsuko also received orders from NHK (Japan Broadcasting Corporation) to improve the shortwave radio receiver, and it was Mr.Jokichi Katagai who had been in charge of that. He would work on the improvements of more than 100 sets of receiver coils in silence. Since I used to make shortwave receivers when I was working part-time as a student, I was very interested in his work, but being a fresh recruit who had just started working, I was disappointed that I couldn't touch the product directly.

I would like to talk about the telecommunication sounder No. 2 that the Ministry of Communications had ordered us to build. This telecommunication sounder No. 2 works like a ringing bell, but has a much clearer sound, and was developed as an efficient sounding device, using a carbon mic instead of a junction. A carbon mic is a part of the telephone transmitter used in phones now.

The company in 1947

Another technology, the power mic (loudspeaker) that was derived from the technology used in this Articulator No. 2, was developed by Iwama-san as another product. It was a little rough, but he applied an ampere-order current to the carbon mic with the intention of producing sound from the speaker without using the amplifier. He was adjusting the speaker every day in the factory, testing the sound. Eventually, it came to be used for announcements on the trains of Japan National Railway (now JR), and whenever I took the Yamanote Line, I could hear the announcement made by the power mic. Then, when election campaigns became a glamorous event, this microphone became popular. This was back when transistors were still not available.

ENCOURAGED BY A FULL-FLEDGED MECHANICAL WORK

Since a short while before I joined, Totsuko had been working on the development of the Hellschreiber telegraphic machine. This is a transmitter of letters developed in Germany, and it breaks one character into 5 x 7 dots to send the data, which is then received by phone or by radio wave, and printed on to a tape using a printer.

Therefore, it is a pair of a transmitter and a receiver. The transmitter was built mostly with mechanical parts, and the main section consisted of three parts, the keyboard section, the alphabet drum section and the junction section. The designer was Mr.Hisashi Inaga, the elder of the Inaga brothers. His younger brother, Katsumi, joined Totsuko about two years later, and designed tape recorders for broadcasting. They were brilliant brother mechanics.

Three months after I joined the company, I was assigned to the production and adjustment of this Hellschreiber telegraphic machine. Now that I was assigned to a full-fledged mechanical work, I was encouraged. The first challenge was to make the junction section disc used to create the letters. I either filed off or left intact the head of the gear according to specifications with the rasp. But how many should I file? I needed as many gears as there were the keys of the typewriter. Which meant more than 76, and making five of them made me sick. So I came up with an idea. I applied my original hand-made rotating jig onto a foot-pedaled shearing machine, and marked the head of the gear which I should leave intact red, and cut off the remaining heads with shears. This allowed me to complete that task in a few days, hardly using the rasp as all. I guess I wasn't the type of person who would think that diligently doing the work that was given to me was good enough. I would immediately start thinking that this task is so tedious, takes a lot of time and inefficient, so maybe I should look for a better way to do it.

As for the receiver, you turn a single helical gear, press the ink roller onto it to wet the tip of the head with ink, run a paper tape along the other side, hit the paper with a plate hammer using an electromagnet, making the helical gear scan the paper sideways and print the characters.

It was Mr.Kaname Nakatsuru who was in charge of this receiver's helical electric printing mechanism. I knew that commercialized pickups and phonomotors were popular from my own experience of buying them in the electronics quarter in Kanda and using them myself. It was Nakatsuru-san who was in charge of producing those pickups. Since then, I worked on development with Nakatsuru-san all through my career. Nakatsuru-san, especially when demand became high for microphones used to record on tape recorders, came to be called the leading expert on the improvement and sound-making of dynamic microphones. Even when I was doing a research on stereo recording, he determined the optimal method for stereo recording from the standpoint of a microphone professional.

The prototype for the Hellschreiber telegraphic machine was completed in September, and we ran a demonstration at the Ministry of Communications, the Central Telegraph Office, the Ministry of Transport and also at Asahi Shimbun. We also ran actual tests at various locations. In particular, the Ministry of Transport was very eager to learn about the performance of long-distance transmission, so we started off with a test between Tokyo and Osaka, then expanded the test area to Nagoya, and then Kanazawa.

MEETING MORITA-SAN

The leader of the demonstration for the Hellschreiber telegraphic machine and its sales was Morita-san. I rarely had the chance to talk with him except for the few times when I saw him in the boardroom. But once the telegraph key facsimile was complete, I started seeing him more often through my work. I remember my first impression about him: he was a very organized person. When running a communication experiment, he would have an experiment plan in his hand, and continue with the experiment in a way such as, "if this result is obtained, let's try this next. If that doesn't work, we'll try this next."

In 1948, we were to run a remote communication test for the Hellschreiber telegraphic machine, and I was sent to Nagoya. This was the first time that I was invited to the main residence of the Morita family located in the Shirakabe district of Nagoya, where I stayed overnight and even met his father Kyuzaemon XIV

(Nobuhide) and his mother. Morita-san took me to his room, and told me about his days when he was a student. One of the closets in Morita-san's room was a tool closet, and it had every tool imaginable hung up on the wall, including a saw, plane, hammer, driver and many more. That was understandable. But what amazed me was that the silhouette of each tool was drawn on the wall. I could see that it made it easy to put the tools away, and it won't go missing. This was another occasion in which I felt that Morita-san was a very organized person.

Morita Shuzo, the sake brewery ran by the Morita family for generations, is located in Kosugaya Village, at the center of the Chita Peninsula in Aichi Prefecture (now part of Tokoname City). I visited there too, and saw for the first time how Japanese rice wine is made. There was a compact car at the factory, a Datsun if I remember correctly. Morita-san pulled it out of the garage to give it a shake down and take a ride in it, which was something he hadn't done in years. He tried to start up the engine, but it wouldn't start. We tried to give it a push-start, but sadly, the engine still wouldn't start up. If it had, it would have been my very first driving experience. Morita-san gave me an extra lesson, and enabled me to broaden my knowledge.

One day, I wondered what this Hellschreiber telegraphic machine was used for overseas, and asked Morita-san. He told me, "it seems that wire service agencies are using it to broadcast market updates." I thought then that maybe I could pick up their signal via short-wave, so I brought my hand-made radio receiver, put up a simple antenna wire and turned the dial to listen in to whatever beat sound came in.

I guess it's strange, but I had an image of what the signals created by the Hellschreiber telegraphic machine would sound like if the radio wave was modulated. The frequency of the turning gear drum and the intermittent rhythm of the junctions were imprinted in my head, so I figured that all I had to do was pick up a station that sounded "trr, trr, trr, trr," from among the myriads of beeping noises. And I found it in less than 15 minutes. Or at least, I thought it sounded right, so I decided to connect it to our Hellschreiber telegraphic machine receiver. Something that looked like letters were printed, but smeared sideways. This was a predictable phenomenon, so I changed the rotating speed of the motor, and then the letters showed up clearly. It looked like some news in English. In all modesty, I assumed that it was not any kind of cryptograph, but couldn't understand any of it because it was all jargon (English classes were nonexistent when I was in college).

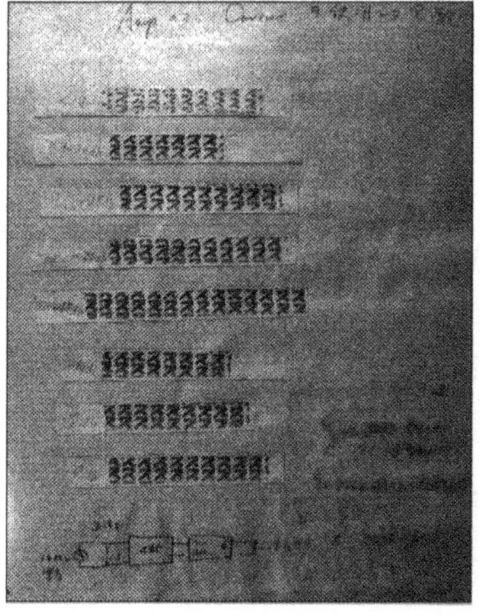

The data for the telegraph key facsimile communication
experiment written by Morita-san

Having found out that I could capture overseas communication, we contacted Asahi Shimbun and I accompanied Ibuka-san and Higuchi-san to run a demonstration. At the newspaper company, we succeeded in receiving signals and printing telegram news from

San Francisco. It was November. I failed to ask if Asahi Shimbun bought our product, but since then, Totsuko and Asahi Shimbun developed a very good relationship.

There was another person who helped me a lot while developing the Hellschreiber telegraphic machine; Mr.Seiichi Nishiyama (who later joined SONY). He used to drop by Totsuko once in a while, but after we started doing telegraphoscopes, he would visit almost every day, and designed a system for the Hellschreiber telegraphic machine that used a communication circuit. It was a system that allowed several telegraphoscopes to be connected to freely communicate between various regions, somewhat like a telephone switchboard using multiple relays. Thanks to Nishiyama-san, the Hellschreiber telegraphic machine were set up in Tokyo, Nagoya and Kanazawa to intercommunicate at will. A practical Hellschreiber telegraphic machine system was completed.

This was how the Hellschreiber telegraphic system was completed by the concerted efforts of so many people, but there were few orders, and I think we eventually gave up on production and sales.

COPYING IS FROWNED UPON

The Tapecorder G-type was handed over to production, and I was relieved of the busy work, but with no time to rest, Ibuka-san told me, "the G-type is way too big. If you make something more portable, we can sell more." So I immediately started putting together ideas to create a new device.

It was a humid and hot night at the end of July. I was in bed, thinking about the device, and new ideas came up one after another. The movement of the levers, the connection with brake operations – all the plans were drawn up on the back of my eyelids, and when I got up happy to have determined that the device would operate perfectly, dawn had broken already. I went to my office early in the morning to draw up the plan in my head before it faded away, and completed the assembly diagram by the end of the day. Whenever I get excited like this, I always forget about time and can't stop working until I'm done.

Also, since I am a machine-oriented person, I have the habit of looking at objects sterically, so I can easily imagine the backside of the object which is not visible in reality, as if I am actually seeing it. So I am very fast at taking the assembly diagram apart and drawing up diagrams for each part, and I never make the mistake of not being able to assemble the parts or the levers not functioning.

I don't know whether it's because I learned Ibuka-san's mindset by osmosis, or because he always made me work on new things, but I never thought about copying American devices, even though I may take a look at them and take hints from them.

Likewise, Ibuka-san never asked me to "make the same thing as this device," however interesting he found the device. He may be attracted to the concept of the device, but it probably never occured to him to copy it. Ibuka-san hates mimicry.

There is another thing that I think Ibuka-san and I had in common. Once something is completed, we are both no longer interested in it. Our interests are already attracted to the next thing, something new.

When the G-type development was completed, he said, "hey, you, the G-type is fine, but can't you make it much smaller? Much lighter? If you can do it, I think we can sell more, at a lower price." He wanted to say that it was enough with the G-type, move on and develop something new quickly.

I had the same idea as Ibuka-san, so I immediately answered, "I will hand over the G-type to production, and start working on the new tape recorder."

Now that I have said that, I must make it happen. And if I am to make it happen, I have to make it something new. If it was a copy of something else, it would be a disgrace. I can never say I made it. So I came up after thinking through the night with a portable device named the H-type.

Locked up in Atami

H stands for Home, which I named with the hope of making a tape recorder for the home. The prototype of this H-type was completed to a satisfactory level within a month. All the know-hows I obtained while struggling with the G-type was put into the newly conceived mechanical section of the H-type, and a very user-friendly device was completed.

Ibuka-san's instinct might have worked. Maybe he thought that this one will be profitable. He must have thought that in order to produce my latest work in the shortest period of time, he should get the best members of the staff together, and prepare the best possible environment to complete the production design as soon as possible.

He decided that the best place to do that would be somewhere with no phone-calls, no visitors, with a good environment – an on-sen (hot spring). Okay, let's go to Atami, he said, and rented a villa in Kinomiya, near Atami, and the two supervisors, Ibuka-san and Morita-san took the designers, Hojo-san, Sekine-san, Inaga-san and myself, as well as Takeuchi-san, an expert material procurement manager, and locked us all up.

Our colleagues envied us, saying it must have been great to go to the spa and draw up plans in yukata cotton robes, but I don't remember feeling comfortable at all, being under surveillance and rushed all the time. In any case, I don't think another lock-up in Atami ever happened again.

CHAPTER THREE

GO THE WHOLE NINE YARDS IF YOU REALLY LOVE IT

what a "life of a tech-head" should be like

1. IT SOUNDS LIKE FUN, SO DO IT RIGHT AWAY

THE H-TYPE APPLIED THE VERY FIRST INDUSTRIAL DESIGN IN JAPAN

It must have been a very edgy idea to lock us up in Atami, but the execution and imagination of the top people at Totsuko was aiming for more than just getting ahead of the times. New and powerful management policies were announced one after another.

First of all, the president of Totsuko changed from Mr.Tamon Maeda (Ibuka-san's father-in-law, a member of the House of Peers and the education minister) to Ibuka-san, and Morita-san became a senior managing director. It was then that the groundwork for Totsuko to develop further was completed.

After that, Mr.Masanobu Tada of Nippon Denki joined us as the leader of research and development, and as my superior, he mentored me in various areas all the way through to the era of video recorders. The book he wrote, "Magnetic Recorder" published in 1953 by Ohmsha, was an excellent book that is read regularly by all engineers all over Japan.

Then, the trademark "Tapecorder" that we had been applying for was finally approved, and publicly announced. Finally, Totsuko's product, the "Tapecorder," not the generic "tape recorder," was to be used publicly. We already had the concept of valuing the brand logo and spreading it and use it widely.

In December, we were to produce 50 samples of the H-type. At this final stage before mass production, the company made an important decision. Back then, Japanese electric appliances were pretty much copies of foreign products, so it was normal for the exterior design to be given little importance. But there were opinions that we should employ a full-scale original industrial design for the H-type.

That is why we asked Mr.Sori Yanagi—a budding designer already famous for his Japonesque and traditional, colorful product designs—to come up with an innovative design that had never been seen before in foreign tape recorders, to produce a trunk-shaped portable recorder as the first industrial design in Japan.

I was very much interested in how my design for a machine would be reborn as an industrial design. Due to operational issues, we held several discussions on how to pass the tape through the head, or what each action meant at each knob position to operate the device, or what kind of shape would best fit the action of the hands.

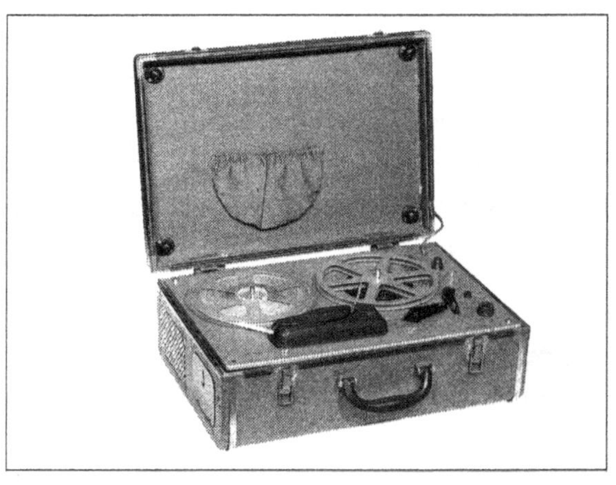

The H-type, which incorporated
the first industrial design in Japan

After the discussions, we decided that the head cover would be a streamlined plastic mold with rounded corners so that it would be easy to insert the tape. The conversion knob was to be a long and rounded spindle-shape, with fine ripples engraved on it to prevent the fingers from slipping. The design was so stylish and meticulous.

The overall form was unique, and though compact, the coloring was very nice too, with the pale green of the panel and the dark brown head cover and knob producing the image of a composed device.

The initial production plan for the H-type was 200 per month, but since we could produce only 100 in reality, the managers apparently decided that the company should put in all efforts into producing the Tapecorder. After that, the pace of production improved, and between June and April of the following year, we were able to produce 3,300 Tapecorders. The market price for one device was 84,000 yen.

In 1951, sales for the Tapecorder soared, while at the same time my research broadened, and I was working on several designs at once. That year, I succeeded in creating a prototype of the "Tapecorder M-type," "Cinecorder (a recording device for movies)," "Tapecorder P (for Portable)-type," "Echo-machine (reverb device)" and "Auto-slide (automatic slide projector with sound)".

THE "PORTABLE RECORDER" WAS MADE USING HAND-WOUND SPRING

I heard that the media was using a portable recorder made by The Stancil-Hoffman, an American company, so I ran around to gather information. With a catalog and some other materials that I obtained, I could analyze its shape and contents.

When management asked me to make a prototype of a portable recorder, I told them that we didn't have any small-sized lead secondary cells and we also couldn't get a hold of an electric governor motor (a rotation-controlled motor). They got me a sample of the battery, so I tested it right away. Though it was small and well-made, complete with a hydrometer, the lead cell always leaked vitriol acid, so I gave up on using it. Instead, I decided to come up with a device that used safer dry-cell batteries.

I also ordered a governor motor, but in Japan back then, technology to produce small-sized motors had not yet developed very well, so I couldn't find something that could be used for the Tapecorder.

I sat down and contemplated how to create a practical method that wasn't just a mimic of American products. Images of parts ran around in my head like a rotating lantern. "Size D batterie", "directly heated tube," "stacked cell," "motor," "governor", "constant-speed rotation", "phonomotor", "portable", "portable phonograph," "hand-wound phonograph", "spring motor"—then I had it. This would definitely work. A constant-speed rotation is possible without the governor motor.

When I was a child, there was a hand-wound phonograph at home that my father oiled and cared for. Watching him, I soon understood that the spring tension was adjusted with a centrifugal governor, and that it was the same mechanism that maintained a train's constant speed using a regulator that looked like a balancing toy on the steam locomotive.

There were many other things that I fixed and maintained with my father, for example, a sewing machine, a camera, 16mm filming equipment, a cinematograph, flat bicycle tires as well as bent and decentralized spokes. We even replacing felt pads on a piano and eliminating squeaking sounds of the piano key joint sections using graphite.

So I was confident that I could make a portable recorder using a hand-wound spring motor. But I had never designed a spring motor. My father's portable phonograph perished in the air-raid, and nobody around me had anything that I could look at for reference.

The following day, I went over to the electronics quarter in Kanda to buy some directly heated tubes to make a battery-driven read/write circuit. When I got to the store that I always go to because they carry a large amount of occupation army surplus goods, I was amazed to find a dozen of the tubes I was looking for, stained with oil, and dumped right there on the ground.

The M-type with its hand-wound spring motor

That store had anything and everything you could want heaped up in piles, such as meters used on airplanes, military-use watches, magnets and servomotors (auto regulatory motors). Most of those would never be used in normal homes. Amidst all of the jumble, my eyes went straight to one particular part that was dumped on the floor. My childhood memories told me that it was a part of the spring used in hand-wound phonographs.

I couldn't believe how lucky I was. I bought two of those and built a test model. Since I didn't have a turntable, I took a flywheel and a capstan (a hoisting mechanism) from a different device to attach to the spring and ran an experiment. It turned out that the tape ran much more stably than I had expected. I showed it to management, got their approval and immediately went out to buy all of the spring parts I could find and created 10 prototypes.

NHK PURCHASED THE PRODUCT

On April 6th, we ran a demonstration of this product to NHK, and received favorable comments, which was a relief to me because the device had so many setbacks. First, it was unable to record continuously for a long time. The spring would come loose after just five minutes.

Second, there was no standard for the absolute tape speed for this device. Crystal servo can be applied easily now, but back then, all we could rely on was mechanical accuracy.

Third, there was the problem of erasing. Due to battery problems, deletion requires 1 watt of power, so I had opted not to add this feature. You had to delete data with a bulk eraser before the next use. You record one use after another, somewhat like the write-once recording method used in CDs now.

The fourth problem was the rewind function. This feature was not a basic essential, but I did prepare something that allowed for a hand-rewind just for the sake of checking test records.

I was anticipating receiving comments like "we won't be able to use such an incomplete product," but since it was very portable and the recording performance was good, NHK deemed that it would be sufficient for professional use.

When it was taken to commercial production, I added a handle-winding feature, and a tuning fork for the strobe as an accessory on the back of the lid to check the tape speed.

"DENSUKE" BECAME THE NICKNAME FOR THE PORTABLE RE-CORDER

The M-1 type used up less battery and hardly required any maintenance, so it was a valuable portable recorder. This was the time when private broadcast stations started receiving approval, and demand for broadcasting tape recorders skyrocketed. This was also a time when Radio broadcasting programs about finding out the word on the street became very popular, so street interviews became a common sight everywhere in town. The M-type performed very well for these recordings. I think I was very lucky to have been able to present the M-type to the market at this time.

Street interviews were something quite new to people in Japan, and since the recorded voice was broadcast later and heard by everyone in the country, people crowded around the interviewer trying to get their voices heard, too.

Mr.Ryuichi Yokoyama, a cartoonist, drew this street interview scene in his four-frame cartoon series "Densuke" that he did for The Mainichi Newspapers. The main character, Densuke, would hold a microphone in one hand with the M-type recorder hanging from his shoulder, running around town giving interviews in a humorous way.

This scene became popular among the broadcasting industry, and ever since, portable devices have been called "Densuke" in the industry.

Since then, the technology used in the M-type continued to improve day by day, and by the time we started producing the M-3, it was running on an electric governor motor, not a spring motor.

Portable devices had one weakness: when recording and playing during transport, the main body shakes, changing the tape speed. With a capstan drive, the miniscule hertz of movement coming from the drive source motor is transmitted to the tape as an irregularity. To eliminate that, I used a flywheel.

Densuke running around taking street interviews

This flywheel is very effective for stationary devices, but if you pick the device up from the table and shake the entire body, the slow pitch of movement can distort the recorded or played sound, as well. This is called a "wow," and with the Densuke-type device, it was not recommended to run around recording sound like the character in the cartoon was doing.

I had been aware of this since I was the one who made the prototype of the portable device, but I was too rushed to take the product into full-scale production and had no time to come up with a proper solution.

IMAGINE THE DEVICE FLOATING IN SPACE

I always built circuits myself, measured them myself and solved even the smallest problems myself. But there is one thing I am confident in: I definitely have a much higher degree of skill than anyone else. This is due to my keen instinct regarding mechanisms.

So then I started thinking, "is there a way to fundamentally prevent wow from being generated? ," for the portable device, I imagined the mechanism floating in a three-dimensional space as usual, and observed inside my head the way the tape run while the device was rocking back and forth.

This device exists in a vast space unrelated to earth. It's not a complicated space like in Einstein's principle of relativity, but rather, the common space of Newtonian dynamics, and it becomes a space of physics, ruled only by the law of inertia.

When the main body of the device is shaking, the tape shakes in the same manner. It is shaking relative to the space. On the other hand, the capstan and the flywheel attached to it try not to move relative to the space, according to the law of inertia. That is why wow is generated by relativity.

Then, I thought, why not control the flywheel in the direction opposite that the main body is shaking in? In other words, I should attach another flywheel that has the same inertial moment in the opposite direction from the main body, and synchronize it so that the wow will be cancelled out.

This was the principal idea of the "anti-rolling method." To make this happen, several additional methods were put into practice, such as using idlers and belts.

Can you understand that you can never achieve this solution if you just stayed on earth and just drew plans on a piece of paper at a desk?

Instantly, I made an improved version of the M-type portable that did not generate wow, and compared it to the conventional portable at a board meeting. I set up a speaker and an amplifier so that the playback sound could be heard by everyone in the room, and holding the carrier belt, I shook it wildly to and fro.

The effect of the anti-rolling was tremendous, and all of the board members were highly impressed. "This is great!" They all asked me question after question about the principle that eliminated the wow, and though I explained it to them, I don't think anyone truly understood. Even if it is common knowledge now, it was probably difficult to realize it for the first time.

I would also like to talk about the Babycorder that was born from the same spatial imagination mentioned in Chapter Three. Incidentally, the professional Tapecorder M-type recorded sales of 2,777 units by October of 1952.

RECEIVING THE "JAPANESE ASSOCIATION OF FILM TECHNOLOGY AWARD"

The Cinecorder is a recorder for cinemas, that is, a recording device for movies. At the time, Japan used optical recording for talkies, but in the West, they were using magnetic recording devices exclusively made for movie recording and editing.

Japanese movie companies such as Shochiku and Toho decided to use this magnetic recorder in their film production, and requested Totsuko to develop a Cinecorder.

Ibuka-san had joined PCL (Photo-Chemical Laboratory), then moved from Nippon Kouon to Nihon Measurement Company. Both PCL and Nippon Kouon were companies in the film industry, so he had many friends who asked him to develop devices used in movie production, and that is why Totsuko decided to develop a Cinecorder.

Again, I was assigned to the development project. I learned the principles of talkies, and studied very hard to learn how the operation flowed in actual studios and how editing was done.

In the West, Cinecorders employed various different mechanisms, such as the RCA method and the Westrex method. We discussed which method we should use, and in the end, decided on the RCA method due to requests from Japanese film companies.

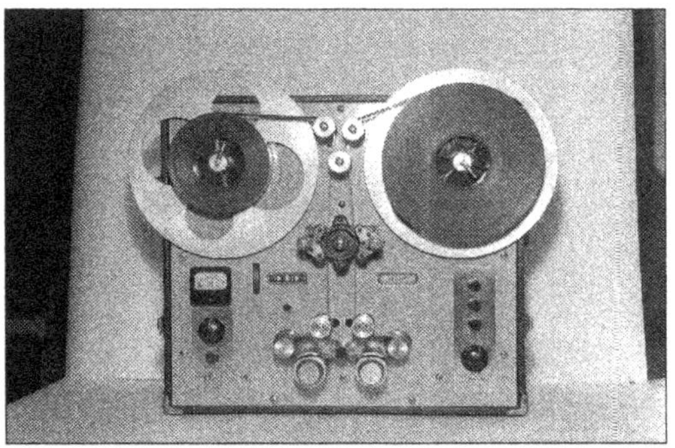

the Cinecorder

The principle of the Cinecorder is the same as regular tape record-ers, except that the tape is a magnetic film, a 35mm film for movies with a magnetic substance applied. Of course, the film has perfora-tions (holes for the gears)for sprockets (gears used to advance the film forward) on either side as with 35mm cinema films. The holes on either side of the 35mm camera film that you are using are the perforations for sprockets.

Magnetic films for Cinecorders use a magnetic substance instead of emulsion, which is where sound for the movies is recorded. The sprocket gear advances the film by gripping the perforation, which makes it possible to match the timing of the sound and the image mechanically.

With the Cinecorder, a completely simultaneous operation without any slips is a must, using the sprocket-driven synchronous motor. Rubber idlers cause slips like with the capstan drive, so it could not be used. You have to use a timing belt, or use the gears to transmit the rotation from the motor to the sprocket. Gear decentralization caused during that transmission, and oscillation generated by the perforation when the gears rub against each other become wows and flutters (sound fluctuation) and degrades the sound quality.

Therefore, many unique driving methods were derived that didn't exist in tape recorders. For example, with the RCA method, one sprocket drives the film in a loop, and sets two rotation stabilizers and the read/write head in the loop.

On the other hand, the Westrex recorder used two sprockets and one stabilizer rotated by the film, with the reading head set inside the stabilizer.

Hitherto unexperienced phenomenon occurred during the Cinecorder mechanism designing stage, and I took great pains in solving these problems, but fortunately, I had Mr.Takeo Tsuchihashi to help me. He has a great deal of experience in film production, and is regarded as a leading expert in optical recording. I asked him to review the history of film production and teach me various information about it.

Tsuchihashi-san is the person who built the first talkie system in Japan, and is known as the inventor of the Tsuchihashi-method talkie (we used to call it the "Dobashi-method", reading the kanji characters of his name differently) whose name always shows up on the title page of Shochiku films. He was the engineer who established the Showa talkie era.

He joined Totsuko as a board member, and became the recording department manager. I heard that all the tools for area meters with optical shutters based on Galvanometers (high-sensitive ammeter) used in optical recording were hand-made by him.

This senior engineer who started the talkie era in Japan was a short man, always easygoing, but hardly talked unless someone talked to him.

He was very kind, and would always say, "hey, if you're going to carry all that stuff, I'll drive you out there," and would always help me take all the filming equipment to Kinuta or Tamagawa. It seemed to me that he really liked cars.

But as for his driving skills…, I was less sure. One day, we loaded the car with all the equipment and started driving. I was in the back of the pickup, so I should have had a clear view behind the car, but a huge cloud of smoke came out of the car and I couldn't see anything.

"I don't know what's going on, but the car seems to be on fire," I shouted at Tsuchihashi-san in the driver's seat.

"Oh, I see. I thought that the car felt slow,"

He said. He had the emergency brakes on all the while that he was driving. Another time, when I was in the passenger seat, he suddenly said,

"There's one part that rusts more than other parts on a car's door. Do you want to know how to prevent it from rusting? "

I didn't even know how to drive back then, so while I was mumbling "Um….," he concluded,

"Well, I was going to tell you something good, but I guess I'm not going to."

There is another funny story. It was raining heavily one weekday morning. I got out of the Gotanda Station to see the road overflowing with water from the Meguro River more than 30cm deep. I rolled up my trousers, took off my shoes and waded my way through to the Mitsui Bank, when I heard a voice. "Hey, Kihara-kun, jump in!"It was Tsuchihashi-san calling from his car. I felt relieved. I had barely finished saying "thank you" when the engine stopped.

I was almost late for work, so I said "I'm sorry" and got out of the car, and waded my way through to the office. But to think back later, it might have been because he stopped the car for me that the engine stopped. His kind intentions might have worked negatively for him. But how did he manage to get to work after that? All these things always make me laugh whenever I think about them.

Thanks to the mentorship of my great seniors, the prototype Totsuko Cinecorder was completed in November 1951, and demonstrated in front of the expers committee of the Motion Picture Engineering Society. We then developed the Cinecorder type-A, type-B and other improved versions, and Cinecorders of various specs went into use.

By December, Toho Cinemas started to produce movies using the Cinecorder, and the subtitle on the title page specified, "recorded by Totsuko Cinecorder."

In 1951, the Motion Picture Engineering Society approved our Cinecorder as a product contributing to film technology, and we were awarded the Motion Picture Engineering Society Award.

UNMATCHING TAPE WIDTHS

By mid-1951, we had started commercial sales of the Tapecorder H-type, and with the M-type gradually penetrating the market, they both were handed over to productions, and I was assigned to work on the Cinecorder and the P-type.

Totsuko had 150 employees then, with enough qualified engineers, and especially with Tada-san becoming the technology manager from May, I was given many subjects to research on, and became very busy.

While doing research for the Cinecorder, I also worked on a new Tapecorder, the P-type, and focused work on its prototype. The main object for developing the P-type was to consider an even cheaper popular version of the H-type. By this time, tape recorders imported from the States had entered the Japanese market, whose specs were not compatible with the H-type, and it was urgent that we develop a new product quickly.

One difference in the specification was the tape width. Tapes for the H-types and G-types were developed originally by Totsuko, so we sold our unique 6 mm spec tape. That was not a problem in the beginning, but eventually, our customers started using imported American tapes in the quest for better sound.

The specification for American tapes was one quarter of an inch, which was slightly wider than that of Totsuko. This could not be run on a Totsuko device. If you tried to force it, the edge of the tape would ride up and damage the tape.

Taking notice of this, we came up with a solution to replace the tape guide to one with a quarter-of-an-inch slits, and to produce all tapes at a quarter-of-an-inch spec.

Then came the second question. The American specs were changed again, and there were to be two tracks recorded on the tape. Until then, sound was recorded using the full width of the tape, meaning it is a single track. You could start recording from one end of the tape all the way to the other end of the tape, where recording would end, and if you rewound it, you could play back about 30 minutes of sound.

With the new spec, the track width was to be cut in half, using the lower bottom of the track one way, and on the way back, flip both tape reels to use the other end of the track to enable twice the amount, so one hour of recording.

The quality of the tape and the head had improved, so even if the track was split in half, the read/ write quality of the device hardly showed any degradation even by the signal-to-noise ratio. This method came to be used in the popular version of our products. It was difficult to switch the H-type to a double-track method, so we started the P-type project in order to develop a common-use version that followed the new specification.

AN OVERARCHING IMPERATIVE: COST CUTTING

Upon analyzing the H-type, I noticed that the cost of the motor stood out. It was the most expensive of all the parts, and its performance was remarkably good for a common-use Tapecorder, so I thought that maybe we should use an induction motor at a level used in home-use phonomotors.

By 1952, power distribution and fluctuations had stabilized, and a 100-voltage fluctuation was not so extreme at night or in the morning. So, I assumed that the induction motor could perform sufficiently in a common-use Tapecorder.

However, the induction motor's torque was small, and though it was enough to run the tape, it was not enough to rewind or fast forward.

The last option that remained, therefore, was the brush motor which had a strong torque. This was the exact same motor used in the model railroad locomotive I played with as a child. The torque was strong enough, but the rotating noise and the sparking electric noise coming from the brush could not be completely eliminated even with insulation shields.

This is how a light and cheap Tapecorder P-1 type capable of long-time recording was completed in response to the demand of that time, with the speed cut down by half to 3. 5 inches. It went on sale in March 1952 at a price of 75,000 yen.

This was 10,000 yen cheaper than the H-type, so in just eight months from May to December, 3,232 units were sold according to company records.

In July of the same year, we released an updated version, the P-2 type. This was a device matching the American spec that used the tape in half-tracks each, and we sold 14,000 of these in one year starting from 1953 January.

2. THE STEREO RECORDING ERA ARRIVES

CREATING A "TALKING SLIDE"

The year 1951 was coming to an end, and it was a busy time of the year. It was decided that the Cinecorder was to be released even though the prototype had barely been completed, so I was working overnight almostdaily to build the prototype and make adjustments. I finally completed it and was relieved to be able to deliver it, when Morita-san came by and asked me to make a talkie slide that coordinated the movement of the slide projector and the Tapecorder.

The slides in this case are not the type where one slide is set in one holder, but rather, like 35mm film, allowed images to be seen as one flowing movie.

I never said no to a new development request. Since this new request involved machinery, I assumed that he came to me because there was no one else he could turn to.

I knew more than a little about the intermittent motion mechanism of the slide's frame-advance feature, since it was used in the toy cinematograph I had when I was a child. By the time Morita-san had finished talking to me, I already had a diagram in my head, thinking that a motor and a gear and a clutch would do the job.

I went out and purchased a commercial projector, upgraded the adaptor section that advanced the film forward so that it would move one frame at a time automatically, and that was all I needed to complete the mechanism of the slide. Next, I decided to use the H-type as the Tapecorder for the Auto-slide, and started working on it.

The principle of the auto-slide is to advanceone frame after another smoothly once the tape starts and the music or the narration begins. By repeating this movement, the story develops.

The most important point is to send the frame forward at the right timing.

There are several ways to do this, but I used the super-practical method at first, pasting a silver foil used in junctions on the back of the tape.

A tape guide adapter is attached on the panel side of the Tapecorder, and the silver foil pasted onto the tape will pass through the guide.

This was a very low-cost method, and the principle was intuitive, easily understood by anyone. Since that was the most important thing, this method was accepted.

An auto-slide, which projects images
in coordination with the Tapecorder

The Auto-slide came to be used in schools as an educational tool, in businesses for promotional purposes, and in a wide range of other uses.

I suppose it was similar to the presentation software that we use on computers now.

Asahi Shimbun wrote about it in the February 13th, 1952 edition as a new product. The article was something like this:

Combining the popular tape recorder with a slide, the prototype of the "talking slide," a hybrid of movies and slides will be released at around April.

This was prototyped by Tokyo Tsushin Kogyo, which attached a type of conductive material like tin foil on the back of the tape to indicate the gap between conversations, and whenever the tape reaches that gap, the electric relay of the slide activates and the slide automatically moves forward.

Slides are much easier and cheaper than producing movies, and in addition, with the sound attached to it, plans to use it in creative ways were already being developed in various fields such as introducing operas, explaining Japanese products to other countries, and explaining about Japan in multiple languages with music. The price for one unit is said to have been a little less than 100,000 Yen.

STARTING THE DEVELOPMENT OF STEREO RECORDING

Ibuka-san visited the United States for the first time in April 1952 and stayed for about three months. During his stay, he decided to obtain a patent license for a transistor, and learned a great deal of beneficial information. Here, I will talk about stereo recording.

Before coming back to Japan, Ibuka-san had the opportunity to hear the sound of a binaural record, and was amazed at the splendid sound heard through both ears. He sent me a telegram from the States letting me know about how "recording with two channels enables a wonderful sound to be replayed."

Hearing that, I quickly decided to upgrade the high-class Tape-corder KP-2 type designated for broadcast stations so that it could record two channels. I used this device when I was doing research on an echo machine, setting the head in two tracks in various positions to extract delayed sound, so fortunately, I didn't have to modify the device much, and by the next day, I had built up a stereo recorder.

Sound picked up by the microphone from both sides is recorded onto two tracks, so ideally, the two heads should be positioned linearly.

That is to say, I wanted to line up the two head gaps, but at the time, the technology to create an inline head had not yet been developed, so I decided to position them in a stagger, and completed the stereo recorder using two microphones.

MY FIRST CABARET EXPERIENCE

Now that I was able to record, I wanted to experiment with how a stereo would sound. I asked Higuchi-san if he or anyone knew anyone in a band, and he said, "the guys in trading should know, so I'll ask them," and he contacted them at once.

That is how I was referred to A-One, a famous cabaret club in Ginza. Nakatsuru-san, who had been developing microphones, and myself carried the equipment into the cabaret club, set up two mic stands, and waited until 8PM when the club opened for business. It was the first time for both of us to be in a cabaret club, and I had no idea what was going to start or what the band was going to play, but after a few songs, I learned the ropes of recording and was even able to enjoy the music while recording it.

This was the first commemorable live stereo recording. The half a dozen band members were playing mainly dance music, and a few couples were happily dancing on the floor.

I was heartily enjoying the live dance music, but as for Nakatsuru-san, he would run up to the microphone every time a song ended, adjusting the distance between the mikes and trying very hard to get the maximum effect of stereo recording.

That night, at around 10PM, we drove quickly back to our office in Gotanda, and hooked up the stereo device to the speaker that we had set up already, and replayed it right away.

THE FIRST PERSON TO HEAR LIVE STEREO RECORDING IN JAPAN

I simply can't explain how beautiful the sound was: a crystal clear resonant sound, separately heard from both sides, allowing you to tell where each instrument is located. The live-ness made me feel a deep emotion that cannot be explained by words.

I ended up sitting on the floor, captivated by the sound, completely losing track of time. I was all the more pleased because I was the one who went through all the trouble of building the device, adjusting it, recording the sound and being met with results that were more than expected.

The ripple effect of this stereo sound is that the sound propagates literally like a wave, and once you experience it, you are captivated by the comforting sound as if it was an addictive drug, and you can no longer live without it.

For decades since then, stereo technology continues to advance unabated, being used in FM radios, Walkmans, CDs and Mini Discs; just about any sound device you could name.

It was Ibuka-san and Morita-san who fell in love with stereo faster than anyone else and worked on its popularization.

As soon as Ibuki-san returned to Japan, he listened to this stereo music and said, "it's even greater than I had expected. The sound that spreads in both directions can be used as a dynamic sound. We must let people hear this." He started inviting people one after another to our office.

As a sample, we recorded moving sounds like a band of musical sandwich-men walking around in the lot in front of the office, or a bike riding around ringing its bell.

Later, we explained all about the stereo to NHK and got to record a performance of the NHK Symphony Orchestra.

With the various sound samples, we explained the reality of stereophonic effects to our clients.

One day, I came up with a little mischievous idea and recorded my own voice on the demo tape. I played a series of sounds by the musical sandwich-men and the NHK orchestra. Then, I walked up from the stereo device that I had been operating, and, getting behind the curtain from the left side of the room, told the clients,

Ibuka-san standing next to the stereo recording speaker

"I'll go to the back of the curtain now, and walk from the left to the right."

So I got back behind the curtain, and touching the curtain, walked to the right side of the room without saying anything.

"In a while, I will reach the center of the curtain…Now I am right in the middle… I am much closer to the right now… and this is the rightmost side of the curtain."

There, I popped out my head from the curtain, and said,

"Could you tell that I was walking from the left to the right behind the curtain? But in fact, I wasn't saying anything. The reason you felt as if the sound was moving was because the sound from the speaker behind this curtain was moving according to the stereophonic effect. How about that? "

The clients had all been thinking that I was talking while moving.

The clients enjoyed this trick very much, and since then, I have to repeat it several times.

GETTING UP ON THE ROOF OF THE HOLY RESURRECTION CATHEDRAL

At around this time, we had come up with various ideas of things to record in stereo, and we had obtained permission for most recordings such as the NHK Symphonic Orchestra concerts' piano, violin, chamber music, and pipe organ. Then one day, Morita-san said, "You know, wouldn't it be wonderful if we could record the bells of the Holy Resurrection Cathedral in stereo? " and so we started preparing to do just that.

I had always had a feeling that Morita-san was a music lover. This was because when I visited his home in Shirakabe, Nagoya when we did the Hellshreiber experiment, I had seen a wide array of records in his room. Most of them were records of classical music and opera collections.

A while after that, when I had a business trip to Nagoya, he said to me, "Kihara-kun, since you're going to Nagoya, can you stop by at my place in Shirakabe and bring all of my records of the opera Cavalleria Rusticana by Pietro Mascagni? " Upon hearing that, I once again thought that Morita-san must really love music.

Such being the case, recording the cathedral bell in stereo was very much the kind of idea that Morita-san would think of. I immediately went over to the Holy Resurrection Cathedral in Ochanomizu, and asked for permission to record the bells. Things went well until I walked right under the bell and saw how tall the edifice was. There was no way we would be able to record the sound using a microphone stand. I asked if we could climb up the belfry tower, and was told that there was no ladder to go up to the bell inside the tower, but there was one outside that would take me to the roof.

So, my partner Shigeharu Iwata and I worked together to climb up the steel ladder, which was about 40cm wide all the way up the vertical wall for about 25m and tied the mic to the pillars on each side of the bell.

At noon, the bell started ringing out its distinctive melody: ding dong, ding dong. We were able to record the sound without incident, but I never want to sttempt such a frightening feat again.

There are two ways to listen to stereo sound: set up more than two speakers, or use a set of headphones. The stereo sound--especially when using headphones--recreates the ambience of the place the sound was recorded, and it sounds so real that you could be tricked into thinking that the sound was actually originating from right where you were.

Back then, our office had a speaker for paging people, and announcements like "Mr.Kato, you have a phone call" would often echo from far away. I recorded this and the surrounding din, such as the conversations of the people in the room, and called Kato-san, saying, "I want you to listen to the latest audio from the stereo recording experiment. Can you come over? ," and he came right away, not knowing that he was about to be tricked. I was such a bad boy!

So Kato-san came over, and I handed him a set of headphones, saying something like, "the new sound is much better." I had another pair of earphones that I used to monitor the sound, and since I knew what Kato-san was hearing, when the "Mr.Kato, you've got a phone call," was replayed, I tapped his shoulder and said aloud, "Hey, Kato-san, you've got a phone call." He quickly removed the headphones and rushed to the phone.

He came back, saying "There weren't any calls. That's funny." I handed him the headphone again without saying anything, and played the same thing over again. This time he realized what was happening and said, "Wow. You really got me. It sounded so real." The audio recording of him being paged that he heard in stereo sounded like it was coming from far away, and seemed so real that it was hard to believe that it was just a recording.

The time signal, "bip, bip, bip, beep" was a sound very well known to people at that time, and whenever we heard it, we would automatically look at our watches.

Quartz watches didn't exist back then. Most watches were balanced and you had to hand-wind the spring using the winder. They were quite inaccurate, and if you forgot to wind them, they would stop. So it was only natural that you would check your watch when you heard the time signal.

My next bit of mischief involved these time signals. "Saito-san, this is the wonderful sound of stereo. Put these headphones on and listen carefully." I said, and replayed the time signal on the stereo tape. The time signal sounded, I watched to see what Saito-san would do, and sure enough, he looked down at his watch, twisted the winder, and adjusted the time. I did it!

"IT'S LIKE BEING BATHED IN MUSIC"

On September 13th, 1952, the newspaper Asahi Shimbun published an article titled "The Era of Stereo Recording? " with the subtitle, "Answer to a decade-long problem".

I will reprint the article here.

The prototype for a "stereo recorder" has been completed. It is a device that uses two microphones to record simultaneously on the tape recorder in order to give a stereoscopic effect to the audio playback. The prototype was completed under directions from President Ibuka of Tokyo Telecommunications Engineering Corporation (Kita-shinagawa 6, Shinagawa-ku, Tokyo), who visited the telecommunications industry in the United States and returned last July after finding some helpful information for stereo recording during his visit.

Conventional recording uses a single microphone, which was like listening with only one ear, but with two microphones, both ears distinguish the sound naturally, which creates a stereoscopic effect and a sense of direction. Sound that enters via two microphones is recorded on one tape each and replayed using two speakers.

Mr.Ryuzaburo Taguchi, Research Institute Staff of Acoustics (at the time), who listened to the sound of an orchestra recorded in stereo, said, "it's like being bathed in music," and commented as follows:

"I attempted this at around 1938, but back then, recording equipment was not so good, and with all the noise, we didn't get the desired results. Anything with huge artistic effects will be recorded in stereo from now on."

This comment by Professor Taguchi, "it's like being bathed in music," I think, is a straightforward way to explain the effect of stereo audio.

Professor Taguchi had previously run experiments on stereo sound back in 1938, so I considered him very much as my predecessor. He mentored me later on acoustic engineering. He was also an authority in the field of color engineering. He helped us again in 1955 with our research on ultrasonic waves, but I will return to that later.

Research on stereo audio had been going on for a long time by then, and I assume that in theory, the technology had already been quite advanced. Japan had not made many achievements in basic research and development of stereo, but I think it was a big breakthrough that Japan was the first in the world to actually use the technology in the broadcasting industry as a new standard. All of this was thanks to Ibuka-san and Morita-san's diligent efforts to let people know about stereo technology.

NHK MAKES THE FIRST STEREO BROADCASTS

On December 4th, 1952, nationwide simultaneous stereo radio broadcast started, possibly the first in the world. This was made possible because NHK had two radio waves, Radio 1 and Radio 2. It was also a result of the bold decisions of Mr.Kazuo Murase, head of NHK facilities back then.

Nakatsuru and I discussed with NHK staff members the idea of editing a 30-minute program of selected pieces from the NHK Symphonic Orchestra that we had recorded previously.

Nakatsuru-san and I discussed with members of NHK's team the idea of creating an edited 30-minute program of selected pieces from the NHK Symphonic Orchestra that we had recorded previously.

After the regular Radio programs ended for the night, and the national anthem was played, our stereo broadcast began.

There was an announcement: "Tokyo Tsushin Kogyo and NHK will now start a test broadcast for stereo." Then, instructions were given." Please have two radios ready. Set one radio on the right to Radio 1, and the radio on the left to Radio 2. Now, you will hear my voice from just Radio 1, so set the volume to the appropriate level.

Next, please set the volume of the radio on the left. This will complete your volume setting. Now, please enjoy the music."

This was how stereo broadcasting started.

We had given advance announcements repeatedly that we would be doing a test broadcast on this day, so many people listened to this night's broadcast, and after the program ended, we received great responses from all over Japan.

Records were shifting from SP to LP at around this time, so there were many hi-fi record fans, and the stereo test broadcast must have stimulated even greater interest in these audiophiles.

We received feedback such as, "I was so impressed that tape recorders could produce such high quality sound," or, "I am amazed that stereo can sound so real." Someone even wrote, "I was listening to the stereo broadcast, and my cat that was sleeping on the table jumped up, surprised by the sound, and ran outside."

By June 1953, NHK had started regular broadcasting of the stereo sound, and the program "Rittai ongakudo (Stereo Music Hall)" was broadcast every Saturday at noon.

Inspired by this, the three main commercial broadcast stations in Japan–Nippon Broadcasting System, Tokyo Broadcasting System, and Nippon Cultural Broadcasting–ran the world's first ternary broadcast.

As such, people in the arts and performers who were constantly in search of new sounds started to skillfully use stereo sound in order to bring new perspectives to movies and theaters.

TERNARY STEREO SOUND USED AT THE HAIYUZA THEATER

1952 was the year of the stereo, and since broadcast stations had begun spreading the buzzword "stereo" on the nationwide network, it became a fad for any business involving sound, industries like movies or theater, to incorporate stereo into their shows and systems, perhaps because people would not come to see them if they did not have the words "stereo" or "stereoscopic" attached to them.

One prime example would be today's new 3-D movies that incorporated images that burst out from the screen and dynamic sound that moves around.

In response to this trend, we cooperated with Toho Cinemas to create audio for sound effects and record ternary stereo sounds, so that the audio would synchronize with the images popping-out of the movie.

This was performed at Nichigeki, a theater located in Yurakucho, Tokyo. Nichigeki later closed down, but at this time, it was a huge music hall that featured movies and line dancing by the Nichigeki Dancing Team.

Recording stereo at Nichigeki (Nihon Gekijo)

The three-dimensional images are watched by audiences wearing polarized glasses, while nine large theater-sized speakers are set up on the left, center and right ends of the theater, allowing sound to move, for example, along with the roller coaster that zipped through the screen during one scene of a movie.

But stereo audio wasn't the only fad of those days. Nearly the entire country was on the verge of a new age of technology. The Korean War that had helped boost the economy was over, Eisenhower had become president in the United States, and since the promulgation of TV had a huge impact on his election, the need for TV broadcasting became much more urgent in Japan, too.

By February of 1953, NHK had begun the first TV broadcasting in Japan, and in August, Nippon Television Network had become the first commercial TV station.

Aside from researching stereo audio and designing equipment, I was personally working on experiments and productions related to television. I would like to talk about that in depth later.

Sometime around May of 1954, I met the composer Mr.Yasushi Akutagawa, who said he wanted to use the stereo effect on a new production he will be playing at Haiyuza Theater, so I lent him the equipment. I met him several times in a month, and had the impression that he was a cheerful and frank person.

While we were talking, I understood that he wanted to have a heavier kind of orchestral music constantly playing in the background while a certain play he had in mind was going on.

I have never been interested in theatre, so I always assumed that only misunderstandings would arise if I talked to artists, but Akutagawa-san was easy to talk with, and he made me feel like I could understand his artistic theories even though I actually couldn't.

"Kihara-san, what kind of music do you like? Oh, piano music. Who is your favorite composer? Shokupan [bread], you say? (a Japanese pun on the name Chopin)? That's a bit too moldy."

This was how he would go on with our conversation." You are an engineer (gijutsu-ya). I am an artist (geijutsu-ya). Anyone who works with 'art' (jutsu) has very good skills, often using them for the good of the people.

By the way, there is something I need very much from you. I hear that you are so smart that you can build just about anything. Can you build me a typewriter that can type musical notes onto a sheet of staff notation? "

Then we talked about composing music. He talked about how hard composition is and that he thought it would help a lot if he had a typewriter for music composition. All I could do was look at the musical score and play the piano, so I said that it is probably impossible to create a device that would type a musical score for orchestras. Akutagawa-san looked very sad. I was sad then too, and feel sorry for him now because if I had a computer back then, I would have been able to grant his wish.

So, our conversation went on like this, and I was kept busy trying to respond to him, but I enjoyed talking to him very much.

In June, we set up a whole system of the stereo recorder in the operation room for lighting at the back of the stage at Haiyuza Theater, and recorded the orchestral performance a few days before the show. Following the script of the play The Red Lamp, Akutagawa-san conducted the orchestra on the stage, and the pieces were performed one after another. After spending more than half a day recording, we made about 10 stereo tapes. Then, I edited these tapes into three completed tapes.

Editing now is definitely dubbing edit, but back then, we had to chop up the tapes and paste them together, so it was a very delicate work to determine where to cut the tapes and to paste them straight. It also took a very long time, and was a great deal of hard work.

By using this tape to play the background music, the theater group didn't have to have a full orchestra play live for them during each performance, which saved them a lot of money; a big advantage for a small theater group like the Haiyuza.

Such renderings at a small theater, which had been impossible until then, we very well received as a novel addition to the experience of theater-going.

MARVELOUS EARS

"Marvelous Ears that Shatter the Industry's Norms" was the title of the article posted in the evening paper of Sankei Jiji Shimbun on December 6th, 1955. It was referring to my ears, saying that they were extraordinary because I could detect ultrasonic waves.

While I was working on the development of Tapecorders and conducting stereo recording experiments, I heard a high pitch tone coming from the TV receiver, I found it annoying, and looked for the source of the tone. Then, I found that the noise was coming from the flyback transformer for high-pressure generation.

I figured out right away that the frequency was a line-synchronizing 15.75kHz, which actually was an ultrasonic wave frequency, supposedly inaudible by human ears. If I was hearing it, I thought that maybe I could hear higher frequencies.

I began turning up the dial meter on the oscillator, and stopped hearing at about 10kHz. That's not right, I thought, and tried listening through different pairs of earphones one after another to test the range of my hearing. I found that the crystal earphones made by RION allowed me to hear up to around 30kHz. It meant that I had become a human tester capable of measuring the features of earphones.

"Marvelous Ears" (with Dr.Taguchi and Ibuka-san)

Ibuka-san caught me running this experiment, and he immediately contacted Dr.Ryuzaburo Taguchi, Laboratory Chief of the Taguchi Psychophysical Research Institute.

We ran a strict test in the listening booth at our office right away, and Dr.Taguchi determined that I was capable of fully hearing up to 29kHz in my right ear and 28kHz in my left.

According to Dr.Taguchi, "the reason that Kihara-kun's ears are so highly sensitive is, since the high tones are assumed to be detected near the entrance of the cochlea around the fenestra ovals, that area of Kihara-kun's ears must have be maintained in particularly good condition."

Ibuka-san said, "Even if we straighten the frequency properties, there is a lot of low quality audio out there in this age of hi-fi. It is desirable that the counterfeit hi-fi rampant in the market now should be purged naturally, but in order for that to happen, we need fans and engineers like Kihara-kun, who can clearly distinguish such high-pitched tones."

Later, I was requested to make an appearance in a popular show on NHK, "My Secret," with announcer Keizo Takahashi as the host of the show.

My secret was that I could hear ultrasonic waves.

The regular panelists were Keishichi Ishiguro (Judoist), Aki Fuji-wara (TV personality, later became members of the Liberal Democratic Party) and some other on-screen talents, but my question must have been very difficult, because no one answered it correctly.

Performing on the TV program "Watashi no Kimitsu (My Secret)"

At the end of the show, a table set up with an oscillator and a speaker was brought out onto the stage, and listening to the sound coming from the receiver, I said, "I can hear it. I can hear it. I can't hear it anymore," at which point the needle on the meter was pointing at exactly 28 kHz.

3. I WANT TO MAKE A "VIDEO RECORDER"

THE DAWNING AGE OF "TELEVISION"

Just around the time that I had completed prototyping the Tape-corder H-type, the Science & Technology Research Laboratories (NHK STRL) located in Setagaya, Tokyo, announced that a test radio wave for television would be transmitted. I was preparing to build a receiver that could receive this radio wave.

NHK had been running open experiments for television image wave reception on a local basis, and had posted descriptions of technical circuits for television on magazines such as "MUSEN TO JIKKEN (The Radio Experimenter's Magazine)" or "The Radio-Craft Monthly." I referred to these resources to work on building a receiver at home whenever I had spare time.

For the picture tube, I used a white phosphor electrostatic deflection tube of 120 mm diameter. It was a monochrome screen optimal for watching television, so I assumed it would be expensive, but when I asked for the price, I was told that it was a defect, so I could have it for free. The defect was that it was out of focus and couldn't be used.

My instincts told me that I could fix the problem, so I took it home with me. I built it into the device, and turned on the raster (dot matrix). It really was totally out of focus, and even though I changed the adjusting voltage within specified values, it couldn't be fixed.

Seeing this phenomenon, I instantly knew the reason this picture tube wasn't functioning properly.

There was some sort of residual magnetism occurring on the iron or the nickel plate used for the electron gun of the picture tube. I eliminated that residual magnetism by putting a "demagnetizer" (used for demagnetizing tape recorder tapes) near the electron gun, and voila, a sharp raster appeared.

I built up the circuit and started making adjustments, but the first problem I ran into was the measurement of the oscillating frequency for high-frequency heterodyne detection (a method to convert received radio waves into a different mid-ranged frequency and amplify them).

So, I hand-made a high-frequency wave detector, and calibrated the range from 100MHz to 150MHz using a short-wave oscillator we had at the office, and inscribed the scale.

This high-frequency wave detector was a very simple one, with a few rounds of thick enamel copper wire wrapped onto a 1cm-diameter bobbin (a frame for winding wires onto) to make a coil, and detected voltage generated on the tuning coil with a small variable condenser attached to it, checking if the needle on the meter moved.

The next problem was the measurement of the mid-range frequency stagger amplifying circuit. I decided to hand-craft the sweep frequency oscillator, too. There are several ways to make this, and the method I chose was to change the coil inductance (a factor that represents how hard a current is to change) in order to change the oscillating frequency, then pass a sawtooth current through it, making the permeability of ferrite (a magnetic ingredient mainly consisting of ferric oxide) change by the strength of the magnetic field.

After obtaining the change in frequency, I could directly see the frequency on the horizontal axis, and the amplification on the vertical axis on my hand-made oscilloscope on a real-time basis as a coordinate, so I was able to make adjustments to the mid-range frequency amplifier in a short amount of time.

The sound was a 4.5MHz FM detection, which was not a problem in theory, but if there was any nonlinearity, there would be buzzing (a spike noise of 50 or 60Hz). It was too bad that the sound was not in stereo at first, but eventually, stereo became the standard.

RECEIVING THE TELEVISION TEST BROADCAST

This is how I hand-made the television receiver, adjusted it, and waited impatiently for the first broadcast to start.

On November 18th (Saturday), 1950, if I remember correctly, the radio wave was transmitted in the morning, and I succeeded in receiving it. All I needed to do was fine-tune the tuning, and I caught the radio wave in no time, and then I adjusted the vertical frequency to get the image.

The program that was being broadcasted was an animated program titled "Kobito to Aomushi (Caterpillar and the Dwarf)," which would be run as a test broadcast every week, and later, a quiz was given on the broadcast.

Receiving NHK's first television test broadcast
with a hand-made TV

For example, NHK would broadcast patterns to measure deflection distortion, and the question would be something like, "answer which pattern was a perfect circle."

Since I succeeded in receiving the images, I took the receiver to work the next Saturday, and showed it to Ibuka-san. He was truly amazed, and said, "You made this on your own? You also made the measurement tools and did the adjustments? This is the first time I saw a TV that received the radio wave. It's got good reception," he complimented me.

After that, he bought me a 17-inch picture tube from the States and also bought me a full-scale electromagnetic deflection TV receiver sample using a 7-inch round picture tube made by an amateur whose name I can't remember.

But later, when other manufacturers continued research on television and started mass production, Totsuko seemed to have little interest in TV for some reason. I suppose it was not that they were not interested in TV, but that they were more focused on the research for transistors that had started in 1953.

THE FIERCE COMPETITION TO BE FIRST IN THE TV BROADCAST

It was the time of power-synchronized broadcasting system, which meant that in Tokyo, the vertical synchronization frequency was 50 Hz, which synchronized to the power frequency. Therefore, inductive interference from the power transformer was not visible, and we had no problem producing receivers.

The test broadcast continued until February 1953 when NHK started a full-scale television broadcast, at which point the system was changed to the NTSC system used now, which, at 60Hz, was the same as the system in the States.

It took a whole year to decide which television system Japan was to use. Exactly a year before, in February 1952, the standard system was established and issued regarding monochrome television broadcast, and the debate between the two systems—the American system using 525 raster, 30 images, 6MHz frequency width suggested by Matsutaro Shoriki, and the system supported by domestic manufacturers using 525 rasters, 27 images, 7MHz—ended with the final decision to utilize the American system.

Now that the television system had been determined, competition to be the first in broadcasting on TV started, with Nippon Television Network Corporation (NTV) obtained the first TV broadcast station preliminary license, while the preliminary license was suspended for NHK and Radio Tokyo.

NHK worked steadily on the preparation, opening the first microwave TV relay broadcast network in Tokyo, Nagoya and Osaka, and started operational tests in preparation for the full license.

NHK obtained its broadcast license in February 1953, and started the first TV broadcast in Japan. The number of television contracts at the end of this month was 1,093 homes.

Six months later, in August, Nippon Television started broadcasting as the first commercial TV station.

The reason NTV started after NHK was because the delivery of the television broadcasting equipment order to RCA in the States was delayed. It was too bad that NTV, ran by Shoriki-san, who had already obtained the preliminary license a year ago, could not be the very first TV station in Japan.

The number of contracts by the end of this month was 3,644, which is a marginal number compared to now, and that year could be called the dawning age of television.

"I WANT TO MAKE A VIDEO RECORDER"

The stereo recording research having slowed down, I was wondering what I should work on next when Ibuka-san brought in something that looked like a catalog, and said to me,

"Hey, here's a catalog of measurement tools by GE (General Electric). There are some frequency generators and oscilloscopes that you might want. Why don't you try making them? "

Watching me create various measurement tools to adjust my hand-made television, he must have thought "this guy could come up with something." Or perhaps, he was beginning to have a desire to sell a high-end measurement tool on the market, which was his goal from when he was in Nihon Sokuteiki (Japan Precision Instrument).

I guess it wasn't a bad idea to build measurement tools, but for me, measurement tools are still nothing more than tools. I am more interested in making things using the tools than making the tools themselves. In addition, electric measurement tools have only a few mechanic parts, so they were not so appealing to a machinery major like me.

I failed to answer him and about two weeks later, Ibuka-san came to me, saying,

"Hey, I got a hold of a GE tube," and he handed me a oscilloscope tube of about three inches in diameter. I was surprised, but I answered, "I'll try to do something with it." And then started thinking about it.

A few days later, I told Ibuka-san, "I'm not quite into developing measurement tools, and it's not in my thing to just copy something that already exists." I then asked him, "So can I make something totally different? For example, I'm thinking that recording television images would be an interesting subject."

That did it.

Ever since then, he never mentioned measurement tools to me. Instead, he said, "Hey, I heard a high-school in Hamamatsu is doing research on an imaging tube called the Vidi-con, so let's go check it out." Ibuka-san and I, along with the freshly recruited Akio Ohgo-shi, went to Hamamatsu High School and learned how to vapor-deposit an amorphous selenium in a vacuum. We also learned how to capture images using that tube.

Using the Vidi-con tube that we obtained, I started making a TV camera. In the meanwhile, Ohgoshi-kun was to work on producing the Vidi-con, and started working with glass.

This was the start of the development of "Video" and "Trini-tron", our two major products in the future.

THE BEGINNING OF MAKING VIDEO FROM CAMERA

In order to develop a video recorder, I needed a signal source. Of course, I had nothing like a TV signal generator. I at least need the same synchronization panel and camera that broadcast stations had.

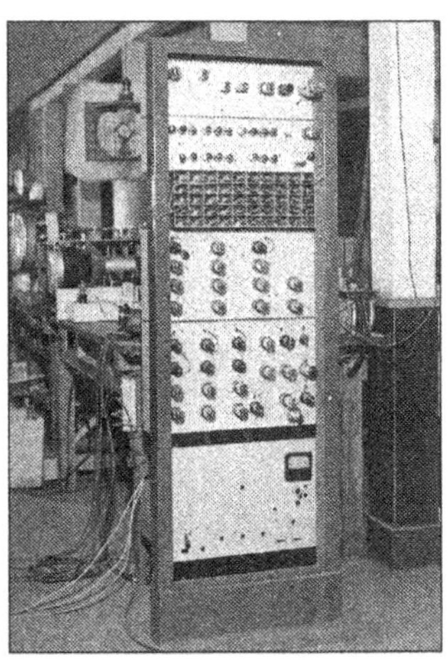

Synchronized signal generator (synchronizaton panel)

I had to start from the synchronization panel. The camera needs a synchronized signal to operate, so I searched for references, but I couldn't find a book that showed a circuit diagram. I asked NHK and was told that I could find it in a film technology magazine called "SMPE (Journal of the Society of Motion Picture Engineers)," so I got a hold of a few issues and read some articles on digital synchronization panels.

The articles explained that a 126kHz pulse from the crystal oscillator was taken and broken down to an eighth, to 15. 75kHz using a flip-flop frequency divider, which was to be the linear synchronization pulse, then counted down 31.5kHz at 525 lines and extracted a 60Hz vertical synchronized signal.

The specified pulse width of the synchronized pulse for the linear synchronized signal and front porch (the leading edge of the signal), back porch (the trailing edge) was set up by selecting a location that was delayed for a designated length of time from the tap (branch) that poked out from the middle of the delay line coil (a coil wound with a signal delaying line). Likewise, specifically designated pulse rows in odd numbers and even numbers for vertical synchronization were created in the same manner.

This basically used the same principles as the synchronized signal generators made easily with IC chips and small crystals now, but back then, I had to make a circuit equivalent to a digital computer, which was not yet known very well, using more than a hundred vacuum tubes.

Thinking back now, I am amazed at myself for lining up flip-flop circuits on my own, working out the actions of the delay line referring to an English book on circuits, measuring pulse width delays with a hand-made oscilloscope and making adjustments. I really am impressed that I did it so well.

Now that I was able generate the synchronized signal, I had to make the TV camera next. This was the first time for me as well, so

I read many books, Japanese and foreign, to study about the subject. TV stations used cameras with image orthicon pickup tubes, and I found many articles on those, but there was nothing on vidi-cons. So, I had to run a cut-and-try test of current values based on ana-logical reasoning for electromagnetic focus and horizontal/ vertical deflection circuits.

Stationary head video recorder

I designed the main body of the camera exactly like RCA's image orthicon camera and decided to attach a single lens rather than a turret (three or four lens attached to a disc that is rotated in order to select the necessary lens). Needless to say, zoom lenses were expensive back then, and even broadcasting cameras used lens-exchange-able turret cameras. So now, I had created a pretty good-looking vidi-con camera, complete with a monitor tube.

REPLAYING ELIZABETH TAYLOR

With the video recorder, I first experimented with a stationary head DC bias recording. Since I used a 5cm diameter capstan di-rectly coupled with the motor for tape speed, the speed was 4 meters per second. Watching the reel, it was turning at a speed that made me think it was quickly rewinding.

I couldn't apply the common knowledge I had accumulated so far on Tapecorders recording at this speed. First of all, I couldn't erase the tape, so I had to erase the entire tape with a bulk eraser before recording. Then I reduced the number of winding on the recording head to about 30 turns, reduced impedance as well so that it would carry current at 500kHz frequency, and set the resonance frequency higher than that.

I measured the recording property in this manner and learned that the limit for replaying reached barely 250kHz. Still, I tried to record video signals from the camera, hanging a calendar photo of Elizabeth Taylor (whom I was very fond of) I set the focus of the video camera, and started recording. I finished recording in about a minute, and replayed it, eager to see the result.

Alas, there was something that seemed like a replayed image, but a horizontal stripe noise covered the screen, and the face of Elizabeth Taylor, which was said to be the most beautiful in the world, had ended up in a decrepit state.

But I could still say that the first recording was a success. After that, I tried changing the head from permalloy to a ferrite head, or adding a high-frequency bias, learning more video recording techniques as I went. Once there was less noise, I started seeing more jitter (a quick shimmering of the images) due to the fluctuating speed of the tapes, so I experimented in every way I could imagine to eliminate the horizontal shimmering on the screen, but this problem remained until the end as an unsolved issue.

POSTPONED DREAMS

Now that I was able to confirm the potential of video recording, I wrote an application for a grant in order to further work on enhancing its performance.

But three months later, I got a reply saying, "your application has been denied, since it is too early for video recording."

On this application form, the project manager was Manager Masanobu Tada, the circuit handled by Masao Karino and the mechanism by me.

The objective of the research was written as, "Tokyo Tsushin Kogyo currently produces and markets magnetic recording tape recorders, and we would like to utilize this technology to produce a high quality recording system for television which will be in demand in the future, with magnetic recording as the technology that will replace the current film recording."

In order to make this happen, one method is to break down the television signal into multiple bandwidths, record the signal from multiple channels to multiple magnetic tracks, and recompose them when replaying.

Another method is to make the relative speed between the tape and the head very fast, and obtain a mega-cycle level of video signal. A method to rotate the head at high speed will be proposed.

The third method would be to propose a method to align the fragmented position–which occurs during high-speed rotation, where the video signal is periodically fragmented when recording–to the position of the synchronized signal, for example, a method to add servo to the rotation, utilizing the feature of the video that displays the images at a predetermined position using synchronized signals. I proposed many other development subjects and applied for other grants, but unfortunately, they were all rejected.

The concept of video recording that I described in the applications were the same methods used in the four-head rotation method that Ampex, a company in the United States, had developed and commercialized a few years later. If we had pursued the video recording development at that time, we might have become the first in developing the video recorder. It was really too bad.

But Totsuko was at a difficult stage back then, with so many investments going into transistor development, the construction of a semiconductor plant as well as cthe construction of a tape plant in Sendai to expand the tape business and increasing capital to fund all of that, so I suppose it couldn't be helped that the development of video recording had to stop.

Incidentally, the idea to read/write using a rotating head had already been used in phonosheets, a method to record sound onto a sheet. Also, Telefunken of Germany already had a patent for the helical scan rotating head before the war. When the video era arrived, these patents came into use. A belated hats-off to the wisdom of our predecessors.

4. THE "TRANSISTOR" ERA ARRIVES

AMATEURS TAKE AN ACTIVE ROLE

When things were in short supply was following the war, there were a lot of amateurs who built radios upon request, myself being one of them. The major electric companies had not yet reestablished themselves completely, so the demand for small things like radios couldn't be met. There were many places where amateurs could take an active roles in things.

The ones who helped meet such demands were the street vendors of electronics in Kanda (now Akihabara), where they had everything from small parts to vacuum tubes in such an abundance that you could find anything you wanted. Many of those street vendors now have huge buildings. In a sense, the electronics industry developed due to the support of these strong dealers, parts manufacturers and subcontractors.

I can never say "no" if someone asks me to do something, so I built a few televisions for people, but that wouldn't be sufficient for me to qualify as an "amateur who took an active role."

There's always someone above you, like one amateur who was a bed seller, but also built cameras for photographs, 8mm cameras and television cameras all from scratch just as a hobby. I read about him in a Jiji Shinpo newspaper article in the fall of 1951.

I showed this article to Ibuka-san and Morita-san right away, and said, "I really would like to meet him and ask him how he builds television cameras," upon which they immediately contacted this man. Morita-san even said, "I was interested in this guy too, so I bought one of the television camera tubes that was in this article and still have it." I was really surprised to hear this.

I checked the RCA vacuum tube catalog and found out that this camera tube was an Iconoscope (model number 5527).

A few days later, I met this amateur, Mr.Yoshio Ozaki, and listened to his war stories about television camera production, which helped me very much.

Ozaki-san, in response to the recruiting by Director Koichi Kasahara and myself, later quit his business of selling beds to join Totsuko. He handled a part of my development tasks, then mainly worked on the upgrade of cameras and telecine, recently, he has put into use the beam recording device for 35mm tapes used for movies, and has also completed digital optical recording for movies, making great contributions at SONY as well.

A Mr.Satoshi Shimada, who was hailed as a child genius, was also working actively as an amateur. His work was so well known that even many people in the television industry knew about him.

He was recruited to our team by Mr.Norio Ohga (later Sony's president and chairman), and developed the small-sized transistor television (micro-TV), spreading the fame and the technical strength of

SONY around the world.

THE "TRANSISTOR" ERA ARRIVES

If we were to say that 1952, the year that Ibuka-san returned from his first visit to the States determined to develop the transistor, was when the project was born, the start of its technology must have been July 1953, when the development team was comprised of Physicists Tetsuo Tsukamoto and Saburo Iwata, mechanical engineer Sukemi Akanabe, chemical expert Akio Amaya, electrical engineer Junichi Yasuda, with Director Iwama at the helm.

By October, Western Electric and Totsuko concluded a contract for technology assistance on transistor production, signed by Morita-san. Continuing research on semiconductors centered around the semiconductor technology from Western Electric, so we started to design a semiconductor production facility.

In October 1954, we sold the first germanium diode and transistor in Japan at the Tokyo Mitsukoshi department store, where we also displayed transistor application products.

Transistor application products included germanium watche, Babycorder, germanium radio (prototype) and hearing aid. There, the 1T23-type diode was sold for 320 yen, and the 2T14-type transistor was sold for 4,000 yen. Other products displayed there included point-contact transistors, junction transistors, power transistors, and photo transistors.

Now that we started working on the transistor, my schedule was overbooked and I was busy working on one project after another.

I had of course started research on applying the transistor to the tape recorder, but Ibuka-san said to me,

"There are thousands of defective transistors that couldn't be used in radios that have been dumped into a huge cask. Can't you do something with them? "

So I came up with the idea of a pendulum clock. A pendulum clock using a junction was already being sold as cuckoo clocks and wall clocks, but I thought that I might be able to create a stable clock using a transistor with no malfunctioning junctions, so I started experimenting. Even if the Beta (current gain) is around 10, it can be used in place of a junction. So, thinking it was a good idea to save the defective transistors, I prototyped a temp-type stand clock as well in addition to the pendulum clock, but it seemed that Ibuka-san didn't like the idea very much, so the product was shelved.

I was also working on a transistor motor at the same time, which was intended to eliminate the electric sparks that were generated from the ultra-compact micro motor brush that became available in the States and in Japan at around 1954. This was the original form of the servo motor, which is now used in Walkmans and video recorders.

DEVELOPING THE "BABYCORDER"

Another thing that I was developing at around this time was the ultra-compact tape recorder. My goal for this product was to read /write using a transistor circuit and to make it as small as possible using a micro motor for the mechanism.

At first, I tried to build a record circuit using the transistor and connected it to the head to supply high-frequency bias from the transmitter, but the output power from the transmitter was too small to obtain an appropriate bias level.

I stared at the circuit diagram to figure out the cause of this power loss, and realized that using the constant current record circuit from the vacuum tube era was inefficient, so I inserted a mixer circuit with a band-path filter into the junction area between the constant current record circuit and the high-frequency oscillating circuit, and made a circuit with no power loss.

As can be said for anything, in order to accomplish something new, you cannot go forward if you are hung up with the old. If you fail, all you have to do is look for a new way.

You many think that you can't find a new way, but it may not be that you can't find it. It could be that you are not trying to find it. If you have the time to think about excuses for why you can't figure things out, then instead just keep on moving forward, even if it is just one step at a time. There will always be a new way.

My next task was to make improvements to the compact DC motor, elsewise I wouldn't be able to make a Babycorder that could fit in the palm of one's hand. The Tapecorder M-3 type "Densuke" for broadcasting that had been manufactured at around this time used a compact DC motor, which was an upgrade from the hand-wound spring motor, but it was still much bigger than the motor I had in mind. In addition, I needed a DC motor to be as efficient as possible, since I wanted to control the rotating speed at a constant rate by using the transistor to add a current servo.

In order to minimize loss in the lateral pressure of the bearing, I used a mini-bearing and altered the pull-out slip-ring at the junction of the centrifugal governor to a point-contact method, and managed to make a motor with minimized friction loss and power loss.

This was the first prototype that was capable of read/write using the transistor in a tape recorder, and was released in October 1954 as a Babycorder using a transistor at an exhibit and sale at Mitsukoshi's flagship store. Although Mitsukoshi had urged us to put this on exhibit, for me, it was still insufficient to call it a complete Babycorder.

It was nothing but a miniature, transistor version of the Tapecorder, with no new ideas put into it. It had nothing that could convince everybody that it was a product that had never existed before.

I suspended development of the Babycorder for a while to wait for a new idea to come to me.

THE FIRST "TRANSISTOR RADIO" GOES ON SALE IN JAPAN

Right after this, I was told to help with transistor radio research, so I immediately joined the radio research group consisting of Karino-san and Yasuda-san to see what they were having problems with. The radio that they had made had a major problem, but it was one dealing with the electric circuit, but rather with manufacturing technology and mechanical design.

The body used a newly introduced plastic mold, which was poorly designed. The form, dubbed "the UN building," had a problem of getting distorted over time due to its thickness, so they had had to give up commercializing the product.

I know how to design machines, but I had no experience with molds. So, I went to a mold factory which was one of our subcontractors, and learned how molds worked to get a clue.

Japan's first transistor radio, TR-55

The corners tend to be easily distorted. The more even the thickness is, the better. If the thickness changes abruptly, the surface tends to shrink. Those were the clues I obtained.

With these clues, I was able to get a rough image of the optimal shape of the radio. As for the arrangement of the circuit, dial mechanism, variable condenser, bar antenna and speaker, I made full use of my hidden artistic senses and put everything together all by myself. I used the downsized mid-range frequency trans, audio trans and speakers that Karino-san had developed, and determined the circuit using my own methods.

While working on this, I got a hold of a 5-crystal radio made by Regency in the United States. I took a look inside and was amazed to learn that they were using a printed circuit board. I had heard about this, but it was the first time I had seen it myself.

Later on, printed wiring would become the norm for circuit boards.

I decided to use the printed circuit board to solder the wiring, so I drew up the diagram for the printed wiring and ordered a subcontractor factory to etch a copper plate, and, finally, I was able to make a printed circuit board. It was April by this time.

Seeing this printed circuit board, Morita-san figured that this would become the mainstream of electric circuits from now on, so he imported all of the material he needed for printed circuits, and started production and marketing.

Printed circuit boards use bakelite sheets with electrolytic copper foils pasted onto them with adhesives, but back then, no adhesives nor electrolytic copper foils appropriate for our purpose were available in Japan, so we had to import them from US R&A (later acquired by Pittsburgh Plate Glass Co.), an American adhesive manufacturer.

Later, a manufacturing company SONY Chemicals Co.,Ltd., which manufactures and markets products such as the Bondmaster and tapes, was established.

In addition to working with Yasuda-san's team, I was also running experiments on my own to make a more compact transistor radio. At this time, vacuum-tube portable radios had been downsized using the POLYVARICON (polyethylene variable capacitor) made by Mitsumi Electric. So, wanting to use that in the transistor radio, I ran some measurements. But I found out that the capacity variation curve of the variable condensers on the high-frequency side and the initiating side was not appropriate for the transistor.

Thinking of consulting the manufacturer to develop an upgrade, I asked Higuchi-san, the plant manager, to set an appointment for me to visit Mitsumi Electric. I met the president Mr.Hajime Moribe, and he agreed to develop a polyester variable condenser for the transistor radio. This was back in February 1955.

Later, Mitsumi Electric made great progress in downsized parts for the transistor and established its position as a parts manufacturer that was essential in supporting Japan electrics industry.

Now that I had gotten a hold of polyvaricon, I started working on the design. For the body, I came up with a shape that minimized distortion. The mold was completed, so I ran experiments like soaking it in hot water and then cooling it rapidly, but there were no recognizable changes in its shape, which was a relief to me.

Having made sure that the body was fine, I attached the polyvaricon and the dial mechanism onto it, along with the speaker, ferrite antenna, printed circuit board and battery box, checked that there were no mechanical defects, and finally, the product was complete.

In April, I handed all of blueprints, circuit diagrams and data to the production team, and the TR-55 left my hands. I have experienced this handover many times and am used to it, but still, it always makes me feel a little sad.

But when Japan's first transistor radio, the TR-55 went on sale on

August 20th for the price of 18,900 yen, I remember feeling very happy about it.

The papers wrote about it too, that Totsuko's technology proved that vacuum tubes could be replaced by transistors, and commented:

Totsuko produced and released the first transistor radio in Japan. The first in the world was Regency in the United States, but since they imported the transistor from Texas Instruments, Totsuko is the first in the world to manufacture everything from scratch, including the transistor. This development of the transistor and the establishment of production technology, as well as the completion of the transistor radio became a benchmark event in the history of our nation's electronics industry. In other words, this shift to the transistor radio proved that it was technically possible for conventional vacuum tubes to be replaced by transistors.

PRODUCING SOUND FROM AN 8 MM FILM

While I was working on the radio, Ibuka-san said to me something interesting. "When you have the time, could you add sound to the 8mm film? "

This all started when Ibuka-san was asked by his friend and physician, Dr.Shigetsune Iikura, whether it was possible for the to "coordinate sound with the tape" of 8mm film, which was in style at the time but had no sound. 35mm and 16mm films could use optical recording for their soundtracks, and by using a magnetic coating, sound could be recorded. Eight mm could use magnetic coating too, but the sound quality was so bad that it was totally impractical and was used only as a toy among amateurs. If we could create a system that coordinated sound with the tape, we would be able to sell more Tapecorders. So, I took on the challenge.

FTS, an 8mm film with sound

But I had absolutely no idea. I was at a loss, and pondered on how I could make it cheap just by using something like an adapter. And as the saying goes, need will have its course. I found a way. The principle is too difficult to explain here, but roughly summed up, it detects the speed difference between the tape recorder and the cinematograph, and controls the voltage of the cinematograph motor.

- A detailed explanation of the principle

Concave and convex grooves are cut into a wheel that is moved by the tape on the tape recorded, a junction is attached, which creates an on/off signal in coordination with the movement of the tape.

Meanwhile, an on/off signal at the same rate is created from the 8mm film drive section, and the two signals are logically combined.

This combined signal is checked for speed, and if the tape recorder is too fast, the motor voltage of the cinematograph is increased to speed up the cinematograph. If it is too slow, the voltage is lowered to reduce the speed of the cinematograph and match the speed to that of the tape.

As a result, both the tape recorder and the cinematograph will continue to turn at the same speed, with the two signals completely aligned, synchronizing in a stable manner.

This method was the precursor of the as the thinking that would later produce the "servo control," and came to be used for 8mm film sound recording in many countries including Germany.

In October 1957, the system was complete, and we invited people from the film industry and journalists to a test screening at a theater in Ginza, and demonstrated the 8mm with added sound using this system. People were greatly interested, and after this demonstration, a few movie magazines asked me to write an article on the system.

On October 11th, Asahi Shimbun posted an article about the system.

▼ Riding the wave of the 8mm fad, a new talkie system combined with the tape recorder was created by Tokyo Tsushin Kogyo. It is called a Film Tape Synchronizer (FTS), an 8mm cinematograph and a tape recorder connected with one cord, synchronizing sound and image so that they produce the sound completely aligned with the image, the sound being much better than the conventional film sound recording system.

▼ Various methods had been experimented with all over the world, but complete synchronization was too costly to be practical. However, this Totsuko system is a simple one costing around 70,000 yen with the tape recorder. Production of 8mm cameras has been soaring lately, with an average of 3,000 units produced each month compared to 1,500 units last year, and is projected to reach 5,000 units by the end of the year. The appearance of this system might also give further boost to its production.

USING THE TRANSISTOR TO MAKE A STROBE LIGHT FLASH

There is a little more to the story between Ibuka-san and Dr.Iiku-ra. The two likely have similar tastes, since one day, they involved me in their discussion.

They were talking was about the model railway, made by German toy manufacturer Märklin. They wanted to make a much more elaborate version of the model railway and sell it as a hobby toy, and they wanted me to help.

I remembered the model railway that I used to play with as a child, but with what they had in mind, I would need to use a transistor to control the train's speed. Dr.Iikura took charge of manufacturing the machinery, and in the end, with Ibuka-san's sponsorship, he launched a full-scale model railway company.

As part of the development of transistor application products, I thought about using transistors in the strobe flash used when taking photographs.

The first output transistor, a PNP alloy type, had been developed for the first time in Japan in January 1955, and a phototransistor had been created six months before that, so I thought I might be able to use these.

Domestic and imported conventional strobes both used a mechanical vibrator, charging voltage boosted from the battery onto the chemical condenser, and trigger voltage to make the xenon discharge tube emit light.

I used the transistor in the part where a DC current of about 200V was generated, boosted using four D-cell batteries, and the circuit part where trigger voltage was generated from the junction of the camera's strobe terminal. Coordinated strobes were quite common by this time, so I had to make a system that made multiple small strobes emit light in coordination with the main strobe. This was where the photo diode came in.

Tsuneo Morita, who had been assisting my work back then, came up with a power-saving method for this strobe system which was only possible with the transistors, and we managed to get it patented.

Once the chemical condenser is fully charged, it stops charging automatically to avoid excess power consumption. Later on, camera manufacturers and strobe manufacturers would be paying us for this patented technology.

Tsuneo-kun worked was very successful in the coming VTR development period.

MAKING AN AUTOPILOT CAR FOR AMUSEMENT PARKS

Totsuko was on good terms with the Mitsukoshi department store, and frequently held exhibitions there. The exhibit and sale in October 1954 was one of them, but in the next exhibition, we scaled things up a bit to promote the technology of Tokyo Tsushin Kogyo, setting up a mock semiconductor plant in which "transistor girls" manufactured transistors.

On the rooftop of many department stores are amusement parks for children, and I came up with an idea of running autopilot cars in which children could ride. It was to demonstrate that this much became possible thanks to our leading edge technology.

There are many ways to run an autopilot engine, but to ensure that it worked and would not fail, I chose to use a servo that could control the steering wheel to run along the white lines of a track. The eyes that look at the white line would be two photo transistors, and worked by feeding back the difference of the white color caught by each transistor to the steering wheel motor; pretty straightforward.

I used a battery-driven DC motor for the engine, and even I, an adult, was able to take a lap, albeit slowly, around the course.

I also made a model carrier ship that runs on a solar battery, which was quite rare at the time.

I set up a couple dozen light bulbs around the water tank to power it instead of relying on sunlight. I pulled the ship with a fishing line so that it would run in a circle in the middle of the tank.

I regret to say that this solar battery was not made by Totsuko. Morita-san had imported it as a sample from the United States. I used it in the carrier ships and transistor radios, and displayed it so that it would be easy to understand the mechanism of semiconductors, which was then unknown technology, that works without a battery.

Building an autopilot car

A ship of about 50cm length that runs on solar battery

INSTALLING A "TRANSISTOR INTERPHONE" AT THE FUKIAGE IMPE-RIAL GARDENS

Ever since Totsuko was established, we had been on friendly terms with the Imperial Household Agency, and whenever we produced something for the first time in Japan or in the world, Their Majesties the Showa Emperor and Empress, the crown prince (current Emperor) and the Imperial family were all very interested, and would come over to our office to look at the products quite frequently.

We presented Japan's first Tapecorder G-type in May 1950 to his Imperial Majesty. We also presented the stereo recorder TC-552 to the crown prince in 1957.

Their Highnesses the Prince and Princess of Mikasanomiya, as well as Their Highnesses the Prince and Princess of Takamatsu-nomiya, visited our auditing room to experience the wonderful quality of stereo in 1955.

I was in charge of operating the machine at that time and answered several questions.

After Totsuko changed its name to SONY, the crown prince would visit the electronics show held at Harumi every year, and for a decade after that, I showed him the big events of that year and gave explanations.

The first year, I showed the crown prince the PV-100 VTR slow-motion replay images, the video sheet recorder the next year and the portable VTR the following year. I had so many honorable occasions to explain our technology to the crown prince.

Based on this relationship, and because SONY was capable of creating the newest kinds of machinery in the world, the Imperial Household Agency asked us to create an interphone that would be easy to use and could connect Their Majesties the Showa Emperor and Empress and the palace staffs. Ibuka-san told me about this, and I started working on it right away. Needless to say, I used the transistor to create a speaker phone that enabled one to communicate just by pressing a button, making it easy to use.

I had the Imperial Household Agency engineering official inspect it and was told to deliver it on June 1961. I drove over to Fukiage in my own car to set it up. I am mentioning my car because when SONY's 15th anniversary ceremony was held in May, Yasuda-san and I received the Superior Employee Award. It was a very honorable occasion, and in addition, we received prize money. It was a huge amount, and I was very surprised when I opened the envelope. I discussed this with my mother and decided to buy a new Toyota Corona, which was delivered on the day I was to go to Fukiage, which is why I drove it over to Fukiage.

I felt like I was gliding over the clouds as I went through the gates of the palace, and having set up the interphone in the bedrooms of Their Majesties, finished everything without issue.

DEVELOPING THE FIRST TAPE MAGAZINE IN JAPAN

Having completed the development of TR-55 radio and after playing around with 8mm films and model railways for a while, I suddenly remembered the Babycorder in July, and tried to remember the ideas I had been thinking about back then. Then I remembered, "Ah, yes. I was going to do something about making it easy to set the tape. There might be an idea for something new in this area."

At around this time, what now is called a compact cassette recorder didn't exist, even in the United States. What we did have was a two-layer magazine reel device, which used up a lot of power with its motor, and the magazine could not be set in one action, all of which were points of dissatisfaction.

I thought up to this point, and then realized that if the reel stage was the reason that the tape reel could not be set onto the reel stage in one action, I could just eliminate the reel stage and rotate the take-up reel using the rotating power of the supply reel. That way, you didn't have to set the reel onto the body of the device and all the device would have to do is run the tape. I attached two reels onto the shaft so that they could rotate freely in the magazine box, and in between the reels, I inserted a spring for balance and made a magazine that was totally independent of the main body.

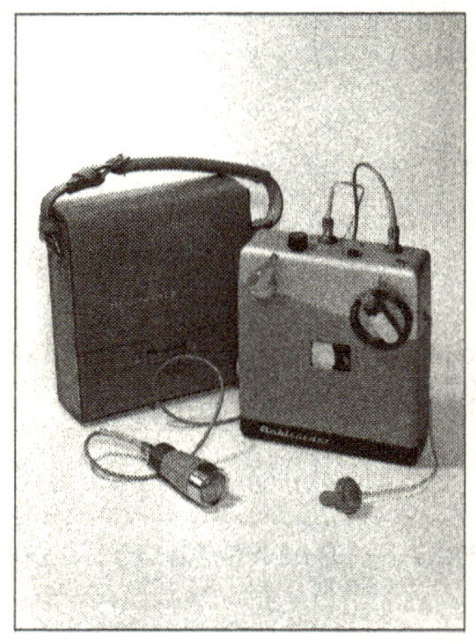

The World's First Transistor Magazine Tape Recorder

This method allowed the tape to be wound onto the reel with a stable tension freely in either direction just by making it run using external power. This magazine method of the Babycorder was the first magazine developed in the world and led to the development of a portable Tapecorder that could read/write with minimum motor power.

LOOKING AT DIAGRAMS BY ROTATING THEM IN THREE-DIMENSIONAL FREE SPACE

I showed the blueprint of this magazine to my boss, who returned it to me saying, "this won't work." Like the anti-rolling method I came up with the portable Tapecorder M-type , if you imagine the device floating in a three-dimensional space, you can get the answer right away. In other words, when you are trying to make something work, there should be no pivot point.

When you are drawing a diagram on paper, parts that are supposed to move around in the device would never be able to move in the diagram. You could try to imagine it moving, but you can only imagine it doing so around the pivot point on the paper. But in reality, that pivot point should be mobile, as well.

In other words, if you draw the diagram in a three-dimensional space, there should be no pivot point. If you happen to set the Babycorder on a desk, anything that is physically fixed to the device cannot move, but everything else should move freely as designed with a mobile axis as a pivot point.

Do you understand? Especially when you are trying to design a device with complicated functions, try to pick up the diagram from the desk and rotate it freely in the air. The device, seen from a different angle to make different movements, should appear in a three-dimensional space.

I completed the Babycorder magazine. It became popular amoung broadcast stations as a transistor portable recorder.

I heard that it was particularly popular among female announcers, since it looked like a purse-type accessory.

[Ultra-compact tape recorder completed: works on transistors that use little current]

This was the title of the article that was posted in the Mainichi Shimbun on August 1st, 1955.

The prototype for our country's first ultra-compact tape recorder was completed by a company in Shinagawa, and expectations for mass production from broadcasting and filming industries are high. This device uses a "transistor," a technology that works like the conventional vacuum tube but is about a tenth its size, allowing it to be much smaller than vacuum tube recorders.

The company had been researching the application of transistors since it obtained a patent from the United States three years ago. The device uses a total of seven transistors for read / write and high-frequency oscillation, with four penlight batteries (6V) attached to its back. The motor has been made to be as thin as possible and is especially small, about the size of a thumb, with the tape holder designed in two layers so that it will not take up space. This small device allows for an amazing ten hours of consecutive recording. Conventional portable recorders type-M weigh 7. 5 kg and its battery had to be replaced every four hours. However, this new device is a solution to all of these problems. In addition, the price of this Babycorder will be about the same as the type-M at 120,000 yen. According to Radio Tokyo's Engineering Department, the mini-phone that they currently use records onto a wire which generates noise, and when the wire is cut off, the connection could be a problem. But this new product uses tapes, which eliminates the problem.

So this was how the Babycorder was handed over to the production team, and went on sale as the BC-1, their first product in April 1956. In October, the upgraded version of the PT-52 went on sale, as well.

DEVELOPING THE "TRANSISTOR TELEVISION"

I was very busy with so many projects to work on in my goal to develop transistor application products. But by the end of 1956, all of my development projects had been completed, and anything that could be mass-produced had been handed over to production, and I finally had time to take a break.

This was when I asked Tada Manager to allow me to work on the development of television, which I had wanted to work on for a long time. He issued me an executive instruction document in November.

I was confident about my knowledge on vacuum tube television, so assuming I could make it by partially replacing the parts to transistor, I started light-heartedly working on the circuit and making a prototype.

But the power transistor was inefficient due to the deflection circuit around the picture tube, and a large volume of current could not be run through it. It was vulnerable to high voltage, and the transistor would be damaged by the pulsive peak voltage, so I had to turn to an imported power transistor.

In order to amplify the image signal, a peak-to-peak voltage of 80V, which is close to the cutoff voltage of the picture tube is required. The transistor withstands only 50V, so this seemed impossible. I performed a push-pull amplification from the previous stage and using two transistors, one used for grid output and the other for cathode output, achieved my goal.

I chose a 25MHz stagger for the mid-range frequency amplification circuit, which could amplify this much of high frequency. I had no trouble with the sound detection amplification.

I knew I was not capable of creating the television tuner from scratch, so I decided to use the vacuum tube for this, and though hybrid, could make a transistor television to a certain level, so I recorded the data and finished my research.

In 1959, the quality of transistors improved, and Shimada-san developed the TV-8-301, which started the era of transistor TV.

CHAPTER FOUR

WHAT THE "SPIRIT OF SONY TECHNOLOGY" IS

"Make Whatever You Need by Yourself"
– This is the Engineer Spirit

1. FIGHTING OUT THE VIDEO WARS

SONY ENTERS THE "AUDIO VIDEO" ERA

As I said before, I was working on primary experiments of fixed-head VTR in 1953. I sent applications to the Ministry of International Trade and Industry for grants for these experiments but was rejected, so I suspended further experiments.

The progress of development of VTR at the time in other countries, as I described in the application forms, were as follows:

(1) Bing Crosby Enterprises – fixed head using the frequency division method, tape speed at 10 in/sec.

(2) Ampex – scan method using a 1-inch width tape and rotating multiple heads

(3) RCA – 0.5-inch width tape, 30 ft/sec, triserial recording track method.

From all these information, we had scrutinized all the methods and had run experiments on anything that was unclear, so I knew all problems regarding VTR.

In 1958, the company name was changed from Tokyo Tsushin Kogyo to SONY Corporation.

In this commemorative year, we pledged to make the next leap forward with a fresh outlook, and a new core product for SONY was born. This was the video.

I think that there were at least three reasons that we managed to develop the video in such a short amount of time.

One was of course our technical strength that we had established in the process of tape recorder development. The next was the voice of authority from Ibuka-san. He was the one who arranged for such excellent people to be invested in the development of videos. The third crucial reason was that we had people who willingly offered us all of the information, tools and data we needed for product development.

The American company Ampex had established their rotating head method and started selling them on April 1957 in the States. But we could not get a hold of any technical information, so it was only when OTV (Osaka TV Broadcasting) introduced an Ampex VTR that we saw the whole picture.

When TBS TV introduced the Ampex VTR as well, I asked Minoru Yoshida Manager to let me go and watch closely how they were recorded and replayed, and I was able to learn a great deal of technical information.

Yoshida-san said, "I would like Japanese companies to develop broadcasting equipment like the VTR as quickly as possible. I hope that the broadcasting business in Japan will grow, and good quality contents would be shown on TV."

I was impressed that he had such a clear vision. Ibuka-san, too, apparently having met Yoshida-san a few days later, said to me, "Yoshida-kun is a bold and big-hearted man. He also seems to work quite actively." Ibuka-san seemed to have fallen in love with Yoshida-san's personality.

Thanks to the technical information on VTR that we obtained, we were able start the developing the VTR. The team was small, being comprised of myself, my subordinate Akinao Horiuchi, Tsuneo Morita and two others, and we started development from April.

Later, the NHK Science & Technical Research Laboratories (NTSL) started a regular meeting titled "VTR Investigation Meeting" where various electric companies gathered to discuss and study measurement results of the VTR.

In August 1958, we managed to get images on the device, so we were finally able to obtain a research grant for that year issued from the Minister of International Trade and Industry for "prototyping research on video tape recorders," and full-scale research on the VTR started.

When we succeeded in producing the Ampex-type VTR, Ibuka-san was overjoyed, and he was also excited to show it to all of the stakeholders.

On December 2nd, we invited everyone who had helped with the research.

Mr.Shigeo Shima, the head of NTSL, visited us on this day. He later became the head of SONY Central Research Institute and then executive managing director and is someone whom I look up to as a mentor.

The following day, I was busy being interviewed by and giving explanations to people from Asahi Shimbun, Asahi Weekly, and others from the MITI.

MAKE ALL "VTRS" USING TRANSISTORS

Now that I had started to understand the concept and the circuits of VTR, I started to think that there was nothing particularly impossible about making it with transistors. It is in my nature to want to try making things right away, so I couldn' t stop myself from asking Ibuka-san to let me work on making the VTR using transistors.

The Ampex-type VTR is quite an elaborate machine using more than 200 vacuum tubes. I figured that if I used transistors, I should be able to make a small device with little malfunction.

I got the go ahead in April 1959, and my six-month challenge with the transistor began.

I made the FM modulator using a flip-flop circuit, the demodulator using a limiter and an integral detection circuit, driving the capstan servo-related motors with a transistor power amplifier, controlling the 60Hz driving frequency using a frequency detection circuit. I added servo to adjust the frequency to 240Hz, so that the rotating head motor would rotate at the specified value of 14,000 per minute.

I also made it possible to record the FM signal going into the rotation head using a semi-power transistor. Since sound recording uses the conventional fixed-head method, I had no problems and was able to delete

the tape using this conventional method. In order to find the optimal circuit, I had to redo the process a few times, but I managed to build a near-satisfactory transistor VTR.

Ibuka-san was delighted again and could hardly wait to release it. So early next year, on January 12th 1960, we decided to hold an exhibit at the Mitsukoshi flagship.

I don' t know if Ampex heard of our exhibit at Mitsukoshi, or if SONY told them about it, but the research managers from Ampex visited us and showed great interest in our transistor VTR. By July of that year, SONY and Ampex had exchanged memorandums on technical assistance regarding VTR.

SONY had interactions with Ampex regarding tape recorders, but after exchanging the memorandums on transistor technology, our relationship grew deeper, with many Ampex engineers visiting us to study eagerly about transistor circuits.

They saw the double-headed VTR with the huge drum that we had been prototyping, and said, "Why are you studying a double-headed VTR? We tried that too, as did RCA. We all gave up on it, so why are you still working on it?" They found it very strange. No matter how much I explained that SONY thinks that the double-headed method and the helical scan have better prospects, they just didn' t understand.

Seeing the huge head drum, they assumed that it was made of a block of brass and joked about it. In the end, they started joking, "Japan must have a lot of brass to waste."

They could not foresee the progress of technology and the development of the times.

But SONY succeeded in building the era of VTR with the helical scan technology. On the other hand, Ampex seems to have fallen behind the era of VTR, possibly because they persisted on using the four-headed method.

THE COLOR PICTURE TUBE FOUND DURING THE FIRST BUSINESS TRIP OVERSEAS

We had started to downsize the VTR, though not because we were mocked for making the head drums with brass. The SV-201 was the all-transistor industrial VTR that we developed first.

Seeing this, Ibuka-san said right away,

"SONY will go with this. Our competitors are racing for broadcasting VTR, and with so few broadcasting stations around the world, the market is small. The market for industrial and home-use VTR is likely to grow from now on, so you should work more on downsizing the VTR."

 In addition, he said,

"Equipment for broadcasting is high-quality and expensive, but probably easy to make. But the home-use equipment has to be smaller and cheaper with the same capabilities, so a lot more knowledge and a spirit of challenge will be necessary. That is all the more reason for us to target home-use VTR."

The first transistor industrial VTR (SV-201) and me explaining it.

This basic mindset and the spirit of challenge is surely what was passed on later as the SONY spirit.

This was the point of origin for SONY, a company that creates a new world without mimicking others and finds new markets by making electric appliances for the home that everybody all over the world can enjoy using.

When Morita-san saw the SV-201, he wanted to show SONY's technical strength. So, he decided to take the latest SONY products to the International Radio and Electronics Show (IRE) held annually in New York.

Transistor radios, various transistors, Esaki diodes, transistor TV, HID tapes and our feature product, the SV-201 VTR, were all packed onto a ship and sent to the SONY America warehouse in New York. This was around March 1961.

Exhibiting the SV-201 VTR at the IRE show in New York

Tsuneo Morita, the one who handled the circuits for the SV VTR, accompanied me. We entered the United States in Hawaii, where we met Mr.Kagawa (the elder brother of SONY's late advisor Yoshinobu Kagawa) at the airport and handed him the NIKON camera that he had asked for. Thinking back now, this could have been regarded as smuggling. But back then, it was normal to be asked by people to transport goods like that.

We visited the SONY America office located at 514 Broadway, which had been established almost a year earlier, and met Mr.Hiroshi Tada and Mr.Kazuya Miyatake for the first time. We held meetings on how to take all of the products to the IRE Show venue and were told about what we should watch out for in New York, which was very handy information back then and even later on.

We completed setting up our exhibit at the IRE Show, made adjustments of the SV, and I ran a demonstration of the device in turns with Tsuneo Morita. The device attracted such a huge crowd that we were surprised. It may have been because it was a rare opportunity for them to see SONY, or perhaps they were interested in the exhibits, but passers-by would say, "SONY. This is SONY," and approach our booth to look at our exhibit with enthusiasm.

While Tsuneo Morita-kun took charge of the demonstration, I had some time to take a look around other booths. Most exhibits of major companies whose names I recognized were demonstrations of color TVs and their applications, while other exhibits consisted of small- and mid-sized parts manufacturers.

RCA had an exhibit of a system that would receive radar information data from air traffic control, convert it into a meshed network of flight tracks with alphabets and numbers, displaying it in color. Since it used a color picture tube, the display was very dark, and the color display was barely visible even though the room was closed off with black curtains.

After that, I walked by the Paramount booth and stopped there. I was astonished by the gleaming brightness of the letters displayed on the color picture tube more than the word processing system they had on exhibit there.

The system was an English word processor that displayed sentences highlighted in red and green. Back then, the only computer I knew of was a huge business-use computer by IBM, so at first, I couldn' t understand the system that displayed what was typed on the keyboard. Unlike a TV scanning method, it seemed like the words were taken in by vector scanning them one by one to be displayed on the monitor. What most surprised me was that the brightness was more than double that of RCA, which used the picture tube in a similar way. The Paramount booth didn' t have the room closed off, and the lights shone high and bright from the ceiling.

Heavens! I ran back to find Morita-san, and almost dragging him, I took him to the Paramount booth to show him. I also took him to the RCA display for comparison

Once Morita-san saw it, it was my turn to be dragged by his speed and ability to take immediate action. He asked about the picture tube to the staff in the booth, but the staff didn' t know very well, so he asked for the name of the president and the address of their New York office, called him right away and made an appointment to see him the next day.

So I accompanied Morita-san the next day to visit Paramount and met the president, and learned that the picture tube was surely a color picture tube using the Chromatron method capable of receiving television images.

Hearing that, Morita-san proceeded to the discussion of contracts for using the Chromatron, and promised that SONY and Paramount would officially negotiate terms at a later date.

Though I knew this was his way of doing things, I was really impressed at the ability to get things done and make decisions so smoothly with the famous film company, Paramount.

The suffering that the Chromatron development team went through was not at all an easy matter.

Yes, the story of "Kurou (meaning "suffering" in Japanese)-matron" is described in SONY history book "Genryu." It really was a blessing for SONY that the Trinitron, invented after all the suffering, was a great success.

THE AMERICANS ARE EASYGOING

There is no end to the argument of positives and negatives regarding characteristics based on people's nationalities. I always try to find the positives in people, so I will talk about two of my experiences overseas.

When I visited Ampex in California, I met with Mr.Markvic, who I befriended in Tokyo, and asked him something I had always wanted to know.

I asked him, "Who decided that the tape guide should be called the 'female guide' ? "The answer was, "It just naturally happened."

Japanese engineers would never say anything aloud that would suggest the delicate parts of a woman. And we would never print it in a user's manual.

The second experience took place at the IRE Show. We had features of the SV VTR itemized and printed on the presentation panel. It said that there were 96 transistors and 35 diodes. People would read it, and smile before walking off. Some people said to us, "this is a good machine." I didn' t understand it at first, but someone from SONY America whispered to me, "it's because there's a 6 and a 9 next to each other," and I finally understood.

Americans love jokes, and humor is natural to them. I was impressed at that.

After this, I visited other countries quite frequently, and experienced at first-hand the difference in the mindset based on nationality. I enjoy talking about these differences whenever they come up.

COMPLETING AND MARKETING THE WORLD'S FIRST INDUSTRIAL VTR

I had the idea of completing the SV VTR as an industrial VTR. In order to obtain the same head speed as the Ampex four-head using a double-headed helical VTR, the diameter of the drum had to be a huge 40cm one. But for industrial purposes, I assumed that using a slightly lower quality at half the speed would be sufficient, so I cut the diameter of the drum to 20cm.

I was experimenting night and day to improve the quality of the SV VTR, and one day, a head made from a "Sendust" alloy, among all of the other various materials that the head development team used, showed unbelievably good performance.

Sendust is an alloy that was discovered by Tohoku University in Sendai, and it is easily turned into dust, hence the name, "Sen-dust." This alloy showed high permeability equivalent to that of permalloy. It was made into lumps using the powder metallurgy method, and was a material optimal for the VTR head with its high abrasion resistance.

Processing was very tough since the alloy is very hard, but we researched polishing techniques and managed to narrow down the head gap to a few microns, 70% of the conventional recording wavelength. We were able to improve the quality of the VTR day by day by continuing to compete on how much the recording wavelength could be cut down in order to lessen the amount of tape used in recording. This later evolved into the era of digital Handycam we see today.

The first step of that progress was the appearance of the Sendust. The heads were measured in a pair, and this "God-head" was held as a benchmark for a very long time. In other words, it took a long time for a head better than this "God-head" to appear.

MAKING VTR USING A 1. 5-HEAD METHOD

Patents for the rotating head method had existed for a long time in many countries, including the alpha-wrap single-head method by Telefunken which was an excellent concept whose basics could be applied to the VTR. The VTR that Mr.Kenichi Sawazaki of Toshiba developed used this alpha-wrap method.

Though the idea of the alpha-wrap was excellent, it had the problem of not being able to replay the vertical synchronization signals in perfect shape, so it was not used for broadcasting purposes. And the new method developed to solve this problem was the 1.5-head method.

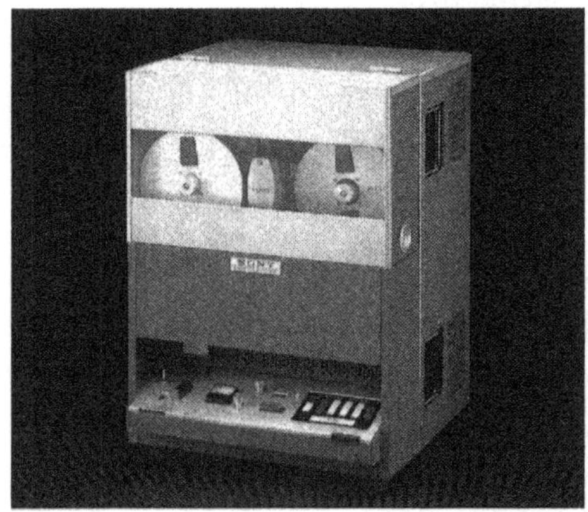

The first VTR to go on sale in the world, "PV-100"

The 1.5-head method uses an omega-wrap on the tape path instead of the alpha-wrap. I will leave the details of this method to specialized books and just introduce the method here.

One head is aligned so that all image signals on the vertical synchronization section will be written and read onto the track onto a tape, and the remaining half head is aligned so that just the vertical synchronization signal can be written and read onto a different track of the tape.

By using this method, downsizing was enabled using a small-sized rotating head drum, and in addition, by using a high-quality Sendust head, the amount of tape used could be reduced by over 30%. This allowed us to develop the PV-100 VTR, a trans-portable that could not be imitated by our competitors. Needless to say, the circuit was all transistors.

In July 1963, we started the sale of PV-100. It was around this time that VTRs were being used frequently in sports events, and in the Tokyo Championship in Athletics, the PV-type VTR was used to determine the order of goals and baton passings in relays, as well as to judge form in gymnastics. When the sixth U.S. -Japan international swimming competition was held, NHK used a specialized PV recording method for its relay broadcast from the States with great success.

The PV-type VTR was also used in an open class in the 14th Annual Conference for Educational Broadcasting and attracted much attention for its excellent educational applications. Thus, it became widely known that the VTR was quite useful in athletics and education, and so the PV-type VTR came to be used overseas for industrial and entertainment purposes. Pan American World Airways used the VTR for in-flight announcements and movies.

Now that the PV-type VTR had become so widely known and used, we started hearing requests from users for much cheaper VTRs.

"KIHARA-KUN IS THE GOOSE THAT LAYS GOLDEN EGGS"

In February 1964, through the good offices of Mr.and Mrs. Higuchi (general manager), I held my wedding banquet with Teiko Goto at the banquet hall of Imperial Hotel Tokyo. Hiroshi Goto,Teiko's father, had worked as the director-general of the The Japan Federation of Employers' Association (Nikkeiren) for many years. When SONY had labor issues back when it was still Totsuko, he helped Ibuka-san and many of the other managers a great deal, and since he knew the people of Totsuko very well, my marriage was arranged.

Mr.Haruo Nagaoka, president of RIKEN (the Institute of Physical and Chemical Research), Mr.Yasuhiro Nakasone, a company representative who later became prime minister, Ibuka-san, Morita-san, general manager Shima-san, general manager Iwama-san, production manager Ohga-san, engineeringdepartment assistant manager Saburo Uemura, second production assistant manager Masahiko Morizono and many other notables attended the ceremony and gave great speeches.

Mr.Tsuneo Morita and Ms. Keiko Kamiya, representing my friends, gave a witty speech which drew much laughter, and I still remember how well they set the mood for the party.

After graduating from the Department of Applied Physics at Waseda University, Teiko worked at the semiconductor research division at RIKEN mainly researching Electroluminescence and reporting her results in academic journals.

When she was in Waseda University, she was the only woman in the department and seemed to have gone through a lot of suffering. But she befriended Kamiya-san who was the first female student in the Department of Science and Engineering, School of Telecommunications at the same university, and they became very good friends. She later became a SONY employee.

The young folks of our engineering department recorded this wedding banquet with their own beloved tape recorder. It was recorded on a tape and a device more than 30 years old now and tucked away inside of a drawer of my desk for a long time, so when I took it out at a much later date, I assumed that there would be a lot of noise on the recording or that the images might be degraded. But when I pulled it out and dubbed it onto an MD, the sound was crystal clear, as if it had been recorded just yesterday. The device and the tapes that I developed were of excellent and reliable quality, for which I secretly praised myself.

I already mentioned that audio that had been recorded onto a paper tape using the Tapecorder G-type was able to be replayed clearly 45 years later. I say again that magnetic recording was a good method. After all of that time, I am very impressed with how well it has stood the test of time.

That aside, when I replayed the recording of Ibuka-san from my wedding 30 years ago, it was like he hadn' t changed at all, except that he sounded a bit younger. His speech went like this:

"If I start talking about Kihara-kun, I can go on for about an hour without boring all of you. Though I will not speak that long, please bear with me for a while.

The first big project that Kihara-kun worked on was the tapes for the tape recorder. He somehow fabricated powder for the tapes by doing something that looked like stir-frying curry powder. Then, he created the tape-recording device almost on his own. It was Kihara-kun as well who made the hand-wound tape recorder Densuke upon my request for a smaller device. Then, a while later, he used the transistor to make a prototype of a television, which was the predecessor of the transistor televisions that we have now. I suppose that was around 1956, and he had already created the antetype of the transistor television.

He is now working on VTR development, and about 20 units have already been exported to the States and put into full use. I see people today recording with 8mm videos, and our dream is to provide them with a VTR as simple as this, available to any family. This is what I want Kihara-kun to develop. I want this dream to come true as soon as possible, which is why I can' t help showing up in Kihara-kun's office every day."

Finally, he added,

"Kihara-kun is the goose that lays golden eggs for our company."

Listening to it now, I am truly amazed at Ibuka-san's foresight. He had ordered me back at this time to develop an 8mm video. I am happy to be able to say that though it took long, I managed to develop 8 mm video just as I had promised.

After that, Morita-san gave the next speech in a very young-sounding voice:

"I have worked with Kihara-kun. There were fun times and lots of bad times too. We often ran experiments blindly, trying all sorts of things, and one day, we were baking some powder and pulled out the thermometer to see the temperature, and the thermometer had melted and stretched like candy. It sounds like a joke now, but back then, we cried over it. Since we have been working together, I have come to think of Kihara-kun as our colleague, and since our colleague was taking forever to find a spouse, we even talked quite frequently about finding him a good wife at company board meetings. So it became one of the company's big projects to find a good wife for Kihara-kun. Now that the project has been accomplished today, I feel like one of the company's plans has been realized, and I am very happy at how it turned out.

I live in New York now, and when I received a call the other day from Tokyo with news about Kihara-kun, I shouted, 'Great. Kihara-kun has a wife now.' My wife overheard that and she too said, 'That's great. Congratulations.' "

This is how he spoke about how happy everyone was about my marriage.

Follow that, Ohga-san gave a speech too:

"Thank you for inviting me today. I have been watching the bride from my seat here and I thought, she sure was worth the wait for Kihara-kun. Now even I feel like I should have waited." (laugh)

Shima-san instantly called out, "Hey, I'll tell your wife you said that."

Hearing this, Ohga-san said, "Oh well. This tape is erasable, so I'll have my part erased later."

Then he continued, "one thing that's missing in all of the previous speeches is that, despite being from the field of machinery, Kihara-kun's knowledge of electronics is top-notch as well. I have heard that President Ibuka taught him electronics at Waseda, but I think

Kihara-kun is better at electronics than even Ibuka-san.

That's why Kihara-kun handles machinery design and electric circuits all by himself. The tape recorder used here was designed by Kihara-kun too. I hope he will continue to develop great machines."

With this wedding, a long life together for the two of us filled with both suffering and enjoyment began. I intend to continue to work on making it an even more fulfilling life.

COMPLETING THE "VIDEO FOR THE HOME"

By having developed the CV-2000, I felt that my initial objectives of downsizing the VTR and reducing its costs were finally achieved.

My promise with Ibuka-san, following his incessantly telling me "It's still too big. Make the next one a tenth this size" was fulfilled by downsizing the PV-100 to a tenth of its predecessor and then down another tenth with the CV-2000; a total of one hundredth the original size, in addition to a significant cost reduction.

I ran experiments on the servo in February 1962. There occured a phenomenon that now seems simple in retrospect, but back then, I would not have been able to explain why it had happened. This phenomenon happened when I was rotating the head drum using a pulley with a rubber belt from the motor. I pinched the head drum axis with my fingers to slow it down and the rotating speed would change quite smoothly.

I was vaguely aware of this phenomenon when I was driving the tape recorder with the rubber belt, but since I had no measurement tools back then, I couldn' t measure the rotating speed in real time, so I overlooked it.

But this time, there was no way I could overlook this phenomenon. If you put a brake on the rotating head axis, the rotating speed will slow down in proportion. I instantly determined that this phenomenon could be applied to the rotating servo.

I reported this braking phenomenon to my boss and the mechanics team, but everyone reacted negatively. They said, "brakes slip, so the movement will be unstable."

I guess that is common sense. Belts used in machine tools and cars transmit horse power, which means that a good deal of friction is generated while they are rotating. There will undeniably be slips, so it is true that the movement is unstable.

So then, why does my brake servo work stably? It is because it operates on a different principle.

The rotating head axis is driven at a speed slightly faster than the specified speed (at 30.3 Hz) by a thin, highly elastic rubber belt from the motor rotating at a certain speed (30 Hz) in synchronization with the power frequency.

The rubber belt on the pulling side of the drive pulley stretches, pulled by the braking force on the brake pulley. The rotating head driven by the rubber belt which has stretched slightly longer than its original length probably rotates slower because the rubber is stretched.

The belt on the other side will shrink after having been stretched. If the belt is stretched a lot, the brake pulley rotation will slow down. The rate it slows down is in proportion to the brake force, so the speed change is smooth and never unstable.

When coming up with new ideas or phenomena, you should trust your own experience and measurements. You should never trust in the common knowledge of other people, or be trapped by your own past experiences.

Unprecedented inventions and discoveries only happen after a breakthrough occurs.

Why was I running this experiment? When I developed the PV-type VTR, Ibuka-san said, "It is still too big. Unless you make both the size and the price a tenth of this, you cannot say it is for home use."

So, I thought about how I might be able to achieve a radical cost reduction. That is how I came up with the brake servo.

DON'T THINK YOU CAN MAKE SOMETHING FROM JUST ONE IDEA

How did I come up with the concept of the Consumer VTR or the CV-type VTR? There should be only one motor, since they are expensive. Avoid usingpower transistors, since they are expensive. Power consumption should be minimized as much as possible. Use as few transistors as possible. In order to do so, use power synchronization for the TV camera. There is no need of a sync disk (synchronizing signal generator). Use a mechanical pulse generator. To cut the amount of tape in half, record using a field skipping method (a method to record the TV screen image using every other frame). In order to do this, use the double-headed helical rotating drum method.

These methods would not be acceptable today. If it was something that could be solved mechanically for example, you were supposed to solve it using machinery design as much as possible and not use expensive electric circuits, so it was unthinkable to use a sync disk. So, I substituted it with a mechanic pulse generator to reduce cost.

In order to apply servo to the rotation head, common sense would tell you to rotate the head motor with an alternating current generated by the power amp and change the frequency of that current. But something new cannot be created without breaking down this kind of common sense.

I decided to try the brake servo that I had experimented with before. For the motor, I chose the cheap hysteresis synchronous motor which was used a lot in home-use tape recorders and rotated the video head drum using a rubber belt stretching from the rotation axis.

This belt servo prototype was a big success, and a single motor could run the video tape, rewind it and rotate the head by applying servo.

Power for the servo worked with 20 V/15 milliampere, so there was no need to use a power transistor in the circuits.

The reason we could sell the CV-2000 VTR at 198,000 yen was not due to just one idea. Various cost reduction ideas and efforts to combine those ideas ultimately gave birth to the world's first home-use VTR.

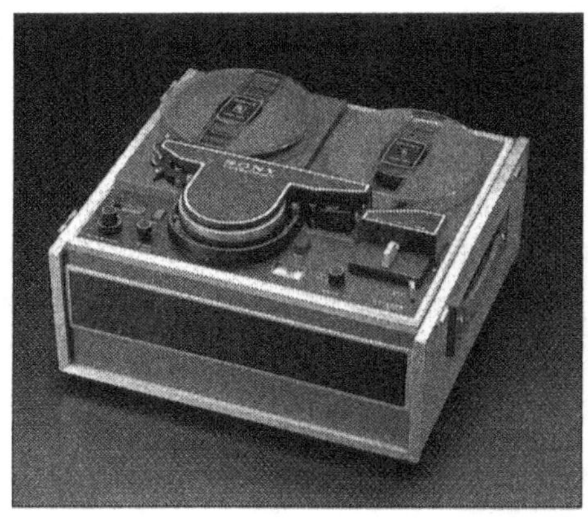

CV-2000 which opened up the dream era of VTR

"VIDEO DENSUKE" IN FULL ACTION IN NEW YORK

In July 1966, we made the world's first portable video tape recorder "Video Densuke" and released it to the public. Newspapers commented about the product: "the Video Densuke is the world's first battery-operated portable video tape recorder which fully leverages its downsized circuit developed for transistor TVs as well as its direct current motor technology and rotation head technology developed for tape recorders. It is sure to make great contributions to broadcasting stations and reportorial assignments."

In August, we held a press conference for the release of the Video Densuke at the SONY office in New York on Fifth Avenue. While Morita-san was talking to the journalists, white smoke erupted from the basement of the SONY showroom across the street, and fire trucks arrived to extinguish the fire. The place was in a panic.

Thinking this was an opportune occasion, I carried the Video Densuke out into the street, and recorded a video the firefighters in action. The cause of the fire was an overheated air conditioner motor , and fortunately, the fire was only a minor one.

I immediately replayed this live video and showed it to the journalists who were all impressed that they were able to see something that had just happened again on video. The press conference was abuzz with this.

A journalist from New York Times reported about this in the morning paper the next day.

"Morita, Sony's president of American operations, was showing off the new device when he looked out the window and spotted smoke billowing out of a neighboring building. He and an assistant ran out of the room and down the street, where they taped the fire and the arrival of the fire department. They then returned to the conference room and screened the tape for the astonished journalist."

He reported this under the title "Trial by Fire Tests New Sony TV Unit."

Kihara filming with the portable VTR,
a device which surprised journalists in New York

RECEIVING THE "SCIENCE AND TECHNOLOGY DISTINGUISHED ACHIEVEMENT AWARD"

With regards to the development of the CV-2000 home-use VTR, I received the Ninth Science and Technology Distinguished Achievement Award in April 1967 for the "single-motor small-sized VTR development".

Looking back now at the development of VTR, I think that the development policies and concepts of the CV-2000 were carried over to the later VTR development with hardly any changes.

Most engineers back then thought that the rotating head method was too complex and expensive for home-use VTR and would never become popular. They likely believed that if they were to try to develop VTR, they would need to use a fixed-head or, in other words, a normal tape recorder method, with an easy and cheaply made device. Based on such thinking, most VTRs used the fixed-head VTR method both in Japan and overseas. Fairchild, Telcan, Newell and RCA from the States, as well as Toshiba, Shibaden and Akai from Japan all released fixed-head VTRs.

The CBS Research Institute in the States also released the "EVR (electronic video recording) method," a method which uses film instead of tape. The whole world was working hard to develop a method for creating home-use video.

I was convinced that the device could be mass-produced cheaply even if it used a rotating head method like the CV-2000. SONY, unlike other companies, is a machinery company as well as an electronics company.

Since it developed as a new form of company, a "mechatronics" company you could say, SONY could always be at the top when it came to the development of new products such as tape recorders and VTRs.

Due to leaders like Ibuka-san and Morita-san who put new ideas into action, the engineers were given a place to work actively, and with everybody enjoying their work and accumulating knowledge on a wide variety of things in a friendly environment, SONY evolved as a company of unrivaled technology.

When I thought about what future VTRs should be capable of, I believed that improving the image quality and downsizing the unit was a must, and I was confident that this could be achieved by usingthe rotating double-headed VTR.

So, as a first step, I created the CV-type. Even when our competitors released products using other methods, I was never swayed and focused on pursuing success with the rotating double-headed method.

Therefore, I never had any excess development costs, which saved both time and manpower. From then on, I continued to release upgrades of the CV-type. I even upgraded the CV-type VTR into a color VTR. My long-cherished Video Densuke became popular in the field.

I also created an all-in-one TV and CV. In Japan, the AV-type became the successor of the CV-type as the standard device designated by EIAJ (Electronic Industries Association of Japan) and came to be used for education and many other purposes all over Japan, and the VTR joined the ranks of "home appliances."

IBUKA-SAN CALLED ME "DIVINE"

Ibuka-san made an appearance on the 18th volume of Nikkei's "Watashi no rirekisho (My Resume)" and talked about the VTR.

"It was around 1962 that we finished Japan's first tape recorder and put it on the market, just when television broadcasting had begun in Japan. Everybody in the engineering department thought that since we could record sound, we should be able to record images on TV. Kihara-kun, who made prototypes for most of the new products made by SONY and contributed greatly to building the foundation of our company, succeeded in making them workafter starting work on them in the summer of 1963.

Kihara-kun is a very skillful and smart man, and if I talk to him briefly about a concept, he will have hand-made something with a much better concept by the next day. It was almost divine to me. I recruited him right after the war since I taught him for a bit when he was studying in the specialized department of Waseda University. He is the one who handled our magnetic tape, all of the early prototypes of the recorders, the first stereo recorder in Japan, the first transistor radio, the first transistor television, the first video recorder in Japan and the downsized transistor version of that. First, he increased the speed of audio tapes gradually at first, finally getting it up to four meters per second. This was a very tough project. Then, he used the same methods from sound recording to record a wavelength up to about 500 kilocycles and managed to get a dim image.

But with this method, a roll of tape capable of recording 30 minutes of sound will be used up in about just 40 seconds. Since this was undesirable, we experimented with various methods including the method to rotate the recording head which is currently being used. We applied for a research grant regarding this issue, but unfortunately, the grant was rejected, and due to several other reasons, this research project was terminated.

If we had continued on with the research, we might have been able to produce the first VTR in the world.

In 1958, an American company called Ampex released a new VTR product, and we rushed to make something similar, which only took us a few months. This is a good example of what I always say, 'just by hearing something is completed, you can make something similar.'

Then, we transformed this into a transistor version, gradually downsized it and now it is much smaller than the vacuum tube TV set. It may still be far from being for 'home-use,' but I am sure it won' t be long before all homes have a color VTR to record TV programs, or record fathers practicing golf in the back yard, then playing it back right away." (Nikkei Shimbun "Watashi no rireki-sho" Vol. 18

What Ibuka-san was thinking about back in 1963 was that, by 1975, everyone in the world would be using Betamax VTR to record programs in color at home, or record their own golf swing and play it back using the slow motion feature. It all came true. I am really amazed at how fast technology progresses.

VIDEOS MUST USE CASSETTES

By this time, the tape recorders used a uniform international standard first set by Philips, so everyone thought that the video should also be a simple device where you just pushed the cassette into the machine.

I thought so too. No one needed to tell me that video players must use cassettes.

It is a reasonably big deal that images could be recorded on tapes, which is 100 times more difficult than recording sound. It might be easy to say video players should use cassettes like tape players, but technically, no one had hardly any experience in that field, so there was no other way to do this save starting from the basics and solving problems one by one.

One of the problems was that, since the video used a helical scan method, the tape runs through the drum and the tape guide diagonally in space. When set in the device, the tape is forcibly pulled to the designated space by the moving guide.

As you may well know, the tape is very thin (about 20 microns). Tape may appear hard when wound onto the reel, but when pulled out of the reel, it is thinner than a piece of paper, so the edges of the tape could be easily folded over or be wrinkled like seaweed when used with tape guides of low quality.

Another problem was figuring out how best to determine the shape of the reel, as well as other formats including size and numbers. Basically, the format is determined by the amount of tape used.

Back then in Europe, a single-reel leader tape method or a two-layered reel method was suggested. Since I always made my designs based on the assumption that technology will progress in the future, when we discussed putting the tape into a cassette, I chose the simplest method, the two-reel cassette without a second thought.

On the other hand, I spent more than six months working on the prototype and brainstorming for ideas for the tape loading method (a method to pull out the tape from the cassette and automatically set it onto the device). I made at least six loading method prototypes in the six months and reviewed each one.

By late 1968, I had used half-inch tape to make the cassette using a loading method called the "drive cast" method, so we released it but didn' t put it into production. This was because the development of "chrome tape" had just come to completionWe had a technical assistance agreement with DuPont of the States, so we were able to use this in our tapes. We had abruptly changed course to start developing the next-generation VTR using chrome tapes.

"Sony succeeds in developing the world's first home-use VTR magazine style color video recorder using the drive cast method." The news spread all over the world, for which the reaction was great, and "Diamond" magazine reported it as the "Magazine VTR," a device one step closer to popularization. The article said,

"SONY recently succeeded in creating a magazine for the VTR which had been an issue common to all companies in the sector. VTR is now one step closer to popularization."

In June 1969, Victor Japan started selling a cartridge VCR. In June, Toshiba released a video tape recorder that could be used both in color or monochrome. In September, RCA in the United States sold their Selector Vision. This used the "digital recording method using hologram technology." In October, Matsushita released a prototype packaged VTR "Magazine method color VTR." This was a year in which we saw one new method after another.

USING THE CHROME TAPE

In 1968, we learned that DuPont had developed a CrO_2 tape, and by August, SONY and DuPont held a meeting regarding the chrome tape.

Chrome tapes have a low Curie point of approximately 120C. When the temperature is raised and cooled back to room temperature, a strange phenomenon occurs. Tapes recorded with little magnetic field increases in residual flux, which means that the tape output increases. Thinking that high-speed and low-cost mass dubbing could be enabled by contact print, DuPont seemed to have been thinking about starting a business that is now known as video rentals, but SONY focused on the advantage of the high density tape, as they wanted to use it in their next-generation VTR.

I happened to visit Philips in Eindhoven, Netherlands in June 1969, and saw DuPont's chrome tapes used in their color VTR prototype LDL-200, and this all assured met that my idea was right.

That is why I decided to used chrome tapes for VTR in the period from 1969 May to October and started to work exclusively on studying circuits that would fully utilize the recording capacity of the tape. This would improve the head and allow it to use as little tape as possible while maintaining quality and fully upgrade the device processing method in order to improve the tracking capability.

With chrome tapes, as opposed to the conventional gamma iron, the squareness ratio of the hysteresis curve is ideal for recording, with the needle crystals thin and uniform in length, and 1. 5 times more saturation flux density, and four times more regeneration output voltage.

When I tested the chrome tape with the CV-type, the regeneration head amp was saturated, and I had to lower the gain for the regeneration circuit.

As a result, the amount of tape used was reduced by a third compared when using the CV-type, and even when we made an hour-long cassette, it didn' t take up a huge amount of space like the open-reel would, which required two 7-inch reels laid side by side. In addition, by setting the tape width at 3/4-inch, the surface area was reduced, and though the cassette became slightly thicker, it became a very well balanced shape for a cassette.

TREMENDOUS IMPROVEMENT IN THE PROCESSING ACCURACY

My primary objective when I was designing the CV-type was to make it cheap. If I had set the processing accuracy any higher than the standard level back then, no factory would have been able to make it for cheap. But after a few years, all industrial machines and plant facilities improved, for example by the use of precision-processing machines such as the NC (numerical control), and mass production became possible.

So, the CV-type was upgraded to the AV-type, and we substituted the field skip method with an all-field recording method. Even if we doubled the processing accuracy and halved the track width, the device remained completely operational, and it was later approved as an EIAJ standard.

In order to improve the tracking quality which would reduce the amount of tape used, a specialized processing machine somewhat similar to a copy grinder was used to improve the processing accuracy of the helical guide, and micron-level accuracy was enabled in the tracking section. If the tracking accuracy is improved, the track width can be made narrower, which would slow down the tape speed and use up less tape.

Progress with the processing technology advanced further, and by the time an automated assembly system using robots was established, we had entered into an era of mass production in which high-precision machines that did not require manpower that were capable of doing things that were impossible for human hands were being widely produced. This was not just about the mechanical section. With IC electric circuits and high integration of parts, we reached the era of Handycam and digital video camera recorders.

TAPE-FRIENDLY LOADING METHOD

I already mentioned how important it is to use a loading method that will not damage the tape. That is nothing special. What I think is the most important thing here is "tape-friendliness," meaning an environment for the tape that allows it to produce the best quality images possible.

A loading method which makes the tapes cry is no good. A forcible tape path that is unkind to the tape will definitely result in the tape crying "squeak, squeak." Can you hear it? Oops, then your video device is a failure. It will not produce acceptable images.

Even if you don' t hear the tapes crying, you should not be satisfied. There may be other factors that could cause the tapes to cry. If you go on designing the device without understanding the causes, problems will persist.

Here, I will discuss how I worked out the tape guide problem. The squeaks that existed from way back in the time of tape recorders were mainly generated by the tape guide. When the back tension of the tape is too strong, or the angle by which the tape winds onto the tape is too big, the tape resonates with the friction of the guide and starts to squeak.

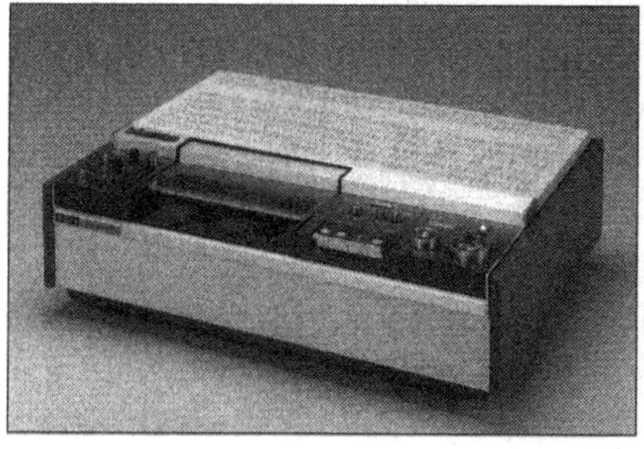

The world's first cassette VTR, "U-matic"

The hand-brakes that I studied in my mechanics class applied brakes by expending little power to generate strong friction. The same principle works on the tape guide, so if you wind the tape onto the guide too loosely, the amount of friction will exponentially increase rapidly. If such a guide exists on the back tension side of the tape, it will amplify the back tension error variation, and extreme jitter will be generated.

Therefore, tape recorders for broadcasting purposes would need an impedance roller (a roller to stabilize the movement of the tape) along the tape to prevent the tape from squeaking.

With the video, the squeaking from the guide and the fluctuating friction at the rotation head drum will be combined, generating a jitter on the screen and seriously deteriorating the quality of the replayed image. But it was possible to design a cassette loading method that eliminated all these problems.

This was the U-loading system, which was used in our product "U-matic." The tape is pulled out by the extraction arm to pass through the deletion head in an almost straight line, wound onto the rotation head drum with no extra force applied onto the tape, passes the sound head to be driven at the capstan, then goes back the way it came through and is wound onto the cassette.

This is how the U-loading system enabled a VTR that could read/write very stable images with little jitter.

USING THE ROTARY TRANS THAT DID NOT GENERATE NOISE

Here, I' d like to explain about the rotary trans which became a secret contributor to the history of videos.

It was August 1958, and I was developing the Ampex-type 4-headed VTR, and saw static on the screen. It might have been due to the power collection rotor and brush attached to the center section of the rotation head having a poor connection. I was bothered with the maintenance of the brush. Then, I thought that the signals could be transmitted by coupling coils instead of using the brush, so I tried wrapping two coils onto the ferrite core and transmitted an FM signal. Then, as expected, the signal was passed on through the air. Then, I thought that the condenser principle might work as well, but this idea failed due to a lack of electrostatic capacity.

Next, I started drawing a plan with the intention of running an operation test with the 4-headed drum using a magnetic method, which turned out to be a very complex plan, and ultimately even I had to give up on it. I had forgotten about this rotary trans for a few years after that, but now and then when people would talk about the noise from the slip ring and the brush in the CV-type, I was reminded that I had to develop the rotary trans for practical use.

The quality of ferrite core materials had improved dramatically by this time, so I made the design and incorporated it into the CV-type, and the insert loss was merely 1.5 decibels. I assumed that this would be acceptable in a commercial product.

So, all VTR after the U-matic was released without the hitherto common noise problems.

ANNOUNCING THE "FIRST YEAR OF THE VIDEO"

With downsized cassettes, the tape loading mechanism can also be downsized in proportion, so I was able to complete a brand new U-loading system that was totally new to the world.

Due to the synergetic effect of all these technologies, the Sony Color Video Cassette System was completed, and it was released in Japan in October 1969 and in New York in November. The appearance of the first genuine video cassette VTR received strong responses.

At this time, the system was called the Sony Color Video Cassette System and didn' t have the name "U-matic" yet.

In 1971, anticipating that the video player would be widely used both in homes and in industrial businesses, thus requiring the sale of a large volume of software, we developed a high-quality master VTR and a video cassette print system to duplicate the needed software and released the "SONY color video cassette integrated system." This video cassette print system is comprised of four cassette recorders in one rack, with several racks operating in parallel, allowing for the dubbing of dozens and even hundreds of cassettes at the same time.

Releasing the color video cassette integrated system
at a private show in New York

In the States, this U-matic VTR became all the rage for industrial and educational purposes, and companies like Coca Cola, IBM and Ford Motors requested the software in bulk, so we set up this dubbing system in print houses of major cities including New York and Chicago to provide various services. This was the beginning of the video rental service which is well known now.

Morita-san proudly announced, "This system marks the start of the era of color videos for the home. This year is the first year for the video age."

SONY then contacted Matsushita and Victor Japan to convince them that the video cassette format should be standardized and announced in December 1969 that "the three companies have agreed to start developing the double-headed helical VTR cassette for industrial use with a tape width of 3/4 inch."

After one long year of waiting, in December 1970, the blanket agreement between the three companies regarding the standardized format of the video cassette was announced.

RELEASING THE "COLOR DEMONSTRATOR"

Having displayed the world's first video sheet recorder (monochrome image magnetic read/write device) and the color demonstrator (color still image player) in 1966 at the SONY Building and the New York Sony show room, Morita-san announced, "we are now capable of offering a new recording media that can be used in demonstrations for industrial or educational purposes."

Systems using magnetic sheets and other applied products continued to be produced, and the floppy disk is now a typical product used in most personal computers. Incidentally, the 3. 5-inch floppy disk is a product that was first released by SONY in 1980 and later became the world standard.

RCA later released a new video disk method as the third version of their Selector Vision. It had a flat surface like an LP record with grooves and pit holes engraved like those of Teledec's products onto which a static pick-up needle was set, extracting video signals converted to electric signals.

Morita-san introducing the color demonstrator

I visited RCA's plant in Indianapolis, Indiana in 1974 and saw their video disc production line. They had several vacuum vapor-depositing devices, and several depositing processes were applied to the discs which would automatically flow through the line to create pressed master discs. I was surprised to see the American way of doing things: spend a large amount of money to complete a huge, full-scale production line even when the product had yet to be released onto the market.

The video record disc is given its conductive property by mixing carbon in the ingredients to extract static, so after being used for a while, carbon deposits would precipitate, with the end of the static pickup instantly being covered with dust and generating dropout noise. They had to solve this problem before releasing the product.

In October 1981, Pioneer Corporation released the "laser disc" at all major electronics and record shops in Japan. Both quality- and price-wise, it was far better than any other method released from other companies, so laser became the mainstream method for discs, and thus the laser application era started.

"U-MATIC" LIVED LONG

It was September 1971 that we started the sale of the U-matic. The color video cassette recorder VO-1700 was sold for 358,000 yen, and the video cassette player VP-1100 at 238,000 yen. The response after their release was very good, especially due to the fact that it was being used mainly for business purposes in the States, with Coca Cola announcing an investment of two million dollars to install SONY video cassettes in 200 of their branches around the country as part of a new educational system for employee training multimedia. In April that year, IBM placed a bulk order of 300 units for marketing and management purposes. In addition, in August, Ford Motors gave us a contract for a bulk order of video cassette players and 60,000 tapes.

The reason that the market expanded so smoothly was because, now that the tape was in a cassette, anyone could use it with ease. It was also a time when videos were recognized as being very effective for educational and marketing purposes.

Another important factor was that the U-matic method was of broadcasting quality. Since the original design left a lot of space for remodeling, it could be remodeled according to broadcasting needs, and by brushing up the circuit, a light and portable system could be set up.

Shortly after the U-matic color video cassette recorder went on sale, CBS, an American broadcasting station, announced the start of their news coverage system using ENG.

ENG stands for Electronics News Gathering, which is a digital news editing system. Conventionally, film recorded on the spot using a 16mm movie camera would be printed and then edited, but the printing took a long time, and also required special equipment and resources.

The U-matic, on the other hand, allowed for great improvements in mobility and time saving. Now, the ability to save time has improved even more, and ENG that use satellite communications have become the mainstream.

Mr.Masahiko Morizono (former Vice President), who offered technical assistance to CBS and completed the U-matic system for the portable ENG, received the Memorial Special Award at the tenth annual Montreux International Television Symposium. The reason for his receiving the award was "for developing the U-matic system for broadcasting purposes and becoming the driving force that resulted in the wide use of ENG."

One method that became practical in the project to develop a color VTR after a few prototypes was the method to add a head for color images in the read/write process. In February 1966, it was announced to the media that a home-use VTR had been developed which would display color images just by adding the color head without changing the basic system of the VTR CV-2000. This product went on sale in the States.

Later, the color adapter CLP-1, developed by Toshihiko Numakura in 1968, was released to the market, but it was not satisfactory as a prospective color recording method for the future.

The next method that Numakura-kun came up with, the "color subcarrier low-pass conversion direct recording method ('low-pass conversion direct recording method' for short)" which was incorporated into the U-matic, and the VTR became a cassette in color.

The patent was published after this, and in December 1970, patents for the concept of the color subcarrier low-pass conversion created by Mr.Mitsuo Fujita (Victor Japan) and Numakura-kun (SONY) were submitted separately. However, since Numakura's patent (published examined application number 49.44535) had been submitted a few days earlier than Fujita's patent (53.9928), the Numakura patent became the standard patent for color.

Having a patent makes a big difference in the competition for development, especially with basic concepts and ideas, so you should try to submit such applications as quickly as possible.

HISTORY OF VIDEO CAMERA DEVELOPMENT AT SONY

It was normal at SONY from the start to make anything you need by yourself. You made a tape recorder, but don' t have a microphone? Then make one. This idea was likely the progenitor to the SONY-ness that exists now.

Likewise, with the video, we had the development of the camera in mind from the beginning. In order to use the camera, we needed a synchronized signal generator, so we decided to make one. What's more, we made it with a digital circuit, which was very new back then. In order to create the best possible system, you must have the energy to make everything on your own. Our enthusiasm was different from the start.

As I mentioned in the chapter on the early ages of television, when I thought about developing the fixed-head video recorder, I immediately figured that I would need a signal source, a TV camera and an Iconoscope camera. Reaching such conclusions in a chain-reaction-like manner, I couldn' t stop myself from running up to Ibuka-san and asking him, "Please buy me a camera tube!"

The great thing about Ibuka-san is that not only does he buy me things like this, but he also think about what's ahead and puts it into action. When I asked him for the camera tube, Ibuka-san searched everywhere, and discovered that the high school in Shizuoka was studying vidicon cameras. He instantly contacted the school and took me and Ohkoshi-kun to visit.

This was the historical event that marked the start of SONY's video camera development. After this, Morita-san bought me an Iconoscope tube from RCA, and I also became acquainted with Ozaki-san, who was known as a video camera amateur, and the video camera development gained momentum.

The early video cameras mostly used Vidicon. We were unable to apply it in a commercial product save for the all-transistor SV-201 video recorder, and used a monochrome vidicon camera.

The PV-100, an industrial use product released after that, also used the vidicon camera, but full-scale usage did not start until the home-use CV-2000 appeared. The CV-type came to be widely used in schools in Japan, so cameras were essential, so we started full-scale mass production of the vidicon cameras.

The CV-type was in high demand in the States too, and the cameras were used in large numbers. But there were many video camera manufacturers in the States, with companies such as GE and Bell & Howell producing vidicon cameras for industrial use.

An OEM deal with these companies came up once, and I ran around to discuss the technicalities with the camera manufacturers. At the time, I had many occasions to visit these companies, which was when I realized that most companies and research institutes had been receiving grants and benefits since they were working with the military or NASA or the government, so though there may be a lot of confidential information, I suspected that they may have already succeeded in expensive technical development and envied them. It is no wonder that the States is leading IC and LSI technology required for military purposes. Their vidicon tubes were of high quality, with good resolution and sensitivity, so I often thought that SONY has much to learn from them.

By this time, the image orthicon tube was used for broadcasting in addition to vidicons. This too was a high-quality, high-sensitivity image pickup tube developed by RCA, and though it was for monochrome broadcasting at first, we used three of these to make a three-tube color camera, and used it for color broadcasting.

DEVELOPING THE WORLD'S FIRST TRANSISTORIZED BROADCAST-ING "IMAGE ORTHICON CAMERA"

In 1964, I thought about developing a monochrome camera that could be used for broadcasting purposes with an image orthicon tube, which would be the world's first all-transistorized camera. So, I had my subordinate, Takao Aoki, work on it. He had experience with vacuum tubes, so he came up with a device that matched the quality of the vacuum tube, or rather, one that was much more stable, much smaller and lighter, and much more mobile. I showed this to some of the people from TBS TV who had helped me on several occasions. They instantly fell in love with this camera and said they wanted to use it for their Tokyo Olympics coverage. We made additional productions to deliver them on time for the Olympic Games.

Back then, broadcasting vans used huge generators, and since the equipment was large too, the car carrying the equipment had to be big. But with the transistorization, the equipment became smaller and used up less power, so a small-sized van was used, and mobility improved greatly. That is why SONY's transistorized image orthicon cameras gathered so much attention.

The following year, the product was handed to the production team for mass production, and more than 50 units of IIC-500 were used in various broadcast stations. In October 1966, it was used by Nippon Television for their live coverage of professional baseball games. I heard that our cameras greatly contributed in communicating the enthusiasm of the final match between the Yomiuri Giants and Sankei Atoms.

The progress of technical development in camera pickup tubes was dramatic, with the appearance of the 3-color camera using three Plumbicon tubes invented by Philips of Netherlands in 1963 with its beautiful images, the image orthicon camera finally disappeared from the market.

Finally, the savior Trinicon tube was made available for home-use color cameras.

DEVELOPING THE "TRINICON SINGLE-TUBE COLOR PICKUP TUBE"

This new color pickup tube was developed at the SONY Research Laboratory led by Director Shima and was released in January 1971 as a revolutionary common-use product of high-quality following the DXC-5000 double-tube color camera. The release said, "the image pickup tube Trinicon uses a single electron beam scanning single-tube method with a built-in tricolor stripe filter and is a revolutionary high-quality common-use color single pickup tube with an optical modulating target and a special electron index element."

Regarding this announcement from Ibuka-san, financial magazines wrote, "SONY showed its confidence at the press release of their compact color camera. On January 27th, in front of the journalists filling SONY's large lecture hall, President Ibuka announced, 'today's release is to announce that SONY is a company which progresses with three elements – camera, video and television – combined. With this compact color camera, we are now poised for that future. ' The release showed his keen motivation to get ahead of the times and incorporate video in the post-color era."

Regarding this announcement from Ibuka-san, financial magazines wrote,

"SONY showed its confidence at the press release of their compact color camera. On January 27th, in front of the journalists filling SONY's large lecture hall, President Ibuka announced, 'today's release is to announce that SONY is a company which progresses with three elements – camera, video and television – combined. With this compact color camera, we are now poised for that future.' The release showed his keen motivation to get ahead of the times and incorporate video in the post-color era."

The research paper written by the five members "Phase Segregation System Single-tube Color Camera by Electron Index," Mr.Yasuharu Kubota, Mr.Hiromichi Kurokawa, Mr.Takashi Shiono, Mr.Susumu Tagawa and Mr.Takehiro Kakizaki of SONY Research Center, was selected and awarded the 1974 Tanba/Takayanagi Award Research Paper Section by the Institute of Television Engineers. Until then, Shima-san, the director of the Research Center, would joke that "we made such a good camera, but no one ever comes to pick it up (tori ni kon)." His frustration on the delayed commercialization of the color camera was clear, but he was likely relieved upon receiving this award.

The U-matic was released for home-use, but contrary to our expectations, it was used for industrial purposes and later came to be utilized as a sub-machine for editing before presenting things on-air at broadcasting stations, contributing to the prime age of ENG. The fact that the U-matic, which is still in use now, was around for an unusually long time compared to most other products is proof of how good its quality must have been.

When I was mainly developing new VTR products as the 2nd Development Division manager at SONY, I had already handed most of the U-system tasks over to productions, so my next project was to be the broadcasting VTR, which is the highest grade of product and something us engineers had wanted to work on for a long time.

Why didn' t we work on it earlier? One big reason was because of a standing agreement with Ampex. In that agreement, it was clearly stated that Ampex will make the broadcasting devices, and SONY will make other industrial VTR devices.

Another reason may have been that Ibuka-san thought that the broadcasting market was small, or perhaps that he didn' t like to interact with public offices very much.

Back around 1970, even when I requested to be allowed to develop broadcasting devices to improve our technical skills, Ibuki-san would say that it was much more difficult to develop complete consumer products than to work on broadcasting products, and that technology for consumer products can eventually be applied to broadcasting products.

The U-system now being used in broadcasting, as well as ENG at its prime, I thought that maybe this was the time for a new broadcasting VTR to come into the world.

Ten years had passed, and the Ampex 4 headed broadcasting VTR was becoming old. I believed that the next-generation VTR would definitely succeed with the helical method, with nothing other than the 1.5 headed method.

Yes. The PV-100, the world's first industrial-use VTR that SONY sold, was the 1.5 head VTR. The PV-type method that I racked my brain to create was a wonderful method that was valid even back then.

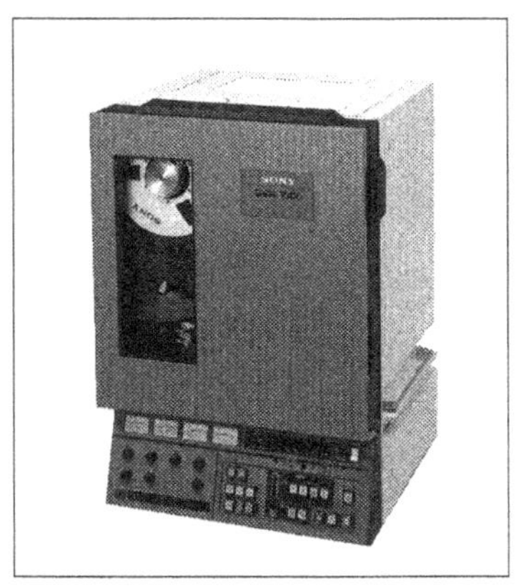

VTR Omega-type

Virtually everything from the materials, parts, tapes, heads, rotary heads, bearings, machine processing accuracy and technology of electric parts and IC circuits that could not be used back when the method was first developed had already reached the level which would enable me to create a broadcasting VTR.

In addition to that, I combined those with the 1.5 head system that I was certain would work and developed the ultimate broadcasting VTR, the Omega VTRs.

In 1975, I ordered Toshio Fujiwara, the leader of the Development Section 2, and a dozen engineers including Shozo Koike to work on this project, and by August, they had made a VTR that complied with the broadcasting standards. After that, I transferred all engineers involved to the Video Section that Morizono-san managed and finally started the production of the broadcasting VTR.

In 1977, the 1 inch helical VTR (1.5 head system) was determined to be the SMPTE standard for broadcasting VTR in the States.

THE DEVELOPMENT GROUP SHOULD START PRODUCTION

This was not the first time that a team of engineers was completely transferred to the production team in order to produce the broadcasting VTR. When I was working on the development of tape recorders, I was all alone, so all I could do was pass along my plans and know-how to the production team. But in the era of VTR, I had many subordinates working under me, so there were groups divided according to projects.

So, like this broadcasting VTR group, the PV-1000 group and the CV-2000 group, as well as the U-matic groups were all incorporated into the production team. For the PV, Inaga-san's group of mechanics transferred to productions. For the CV, it was Mr.Sadayoshi Kato. For the U-matic, it was Mr.Akinao Horiuchi. Each time, the project leader along with one or two dozen engineers would transfer to productions.

Higuchi-san, who has known me for a very long time, once teased me, saying,

"The good thing about you is that you never interfere with what the production team does with the product you developed. You are quite easy-going about it. Everyone is happy that you leave the job up to them. Some people can get really nosey and be annoying."

I never bother my subordinates with micromanagement. The more brilliant they are, the more I would rather have them work elsewhere to exert their full potential. Producing products they developed on their own saves time, and even if a problem arises, the engineers who developed the product will be able to solve it right away. What's more, they have an attachment to what they have developed, so the product will be produced with the greatest possible care.

2. THE DIFFERENCE BETWEEN THE MINDSET OF WESTERNERS AND JAPANESE

WESTERNERS RESPECT ACHIEVEMENT OF OTHERS

Starting in 1961, I visited various countries in Europe. I had many occasions to have relaxed discussions with people who worked in development at business dinners and parties. To summarize what I felt during those occasions, I feel that the Westerners have a mindset of "respecting others and never doing something others don' t want." I am sure that there are people who do not have this mindset as well.

But people who have been educated well are extremely courteous. I personally think that it is their devotion to Christianity that makes such people think "whatever I don' t like, others don' t like either. So I will not do those things to others."

For example, there are many ventures in the States that produce strong results based on wonderful ideas. Most of the time, major companies rarely try to forcibly crush these small ventures to take advantage of their ideas and hard work. They respect the ideas of others, so they would never try to copy or steal from smaller ventures.

They know to acknowledge the achievement of others and respect them, and are conversely they are extremely contemptuous about those who break the rules.

Perhaps for this reason, it seems that there is only one major company in each field when it comes to the manufacturing industry in the West. Needless to say, the service and entertainment industries are not manufacturers, so their behavioral patterns are likely totally different.

Some companies are even synonymous with their respective industries, for example: Kodak for films, Xerox for copying machines, 3M for tapes, IBM for computers, DuPont for petrochemicals, Polaroid for instant cameras, Zenith for color TV and Philips for home electronics in Europe.

It is said that in Japan, there is no setting in place for ventures to grow. There are no big-name companies or millionaires who are willing to help small ventures. Rather, people rush to ideas that they are sure will make money, and start shamelessly copying from them.

I don' t ascribe to the idea that anything is okay as long as you are not breaking the rules. There should be issues of thoughts and implicit rules about respecting the great achievement of others. I believe that clean relationships and clean businesses is only made possible when we acknowledge and understand one another. I had a few occasions to meet Dr.Edwin H. Land, the founder of Polaroid. The first time was in 1957, when I visited their research center in Boston thanks to a reference from Morita-sanI observed their experiments with their newly developed high-sensitive instant films and listened to the explanation of the dichroic method invented by

Dr.Land.

The principle of the dichroic method as applies to color TVs is to film the image with only two picture tubes, one for monochrome and the other for red. When the two signals are entered into the color picture tube of just two colors of white phosphor and red phosphor, a color image is displayed.

Listening to the explanation, I thought, "oh, I got it." I remembered that when I had the two different colored light bulbs on, the shadow of my hand looked blue. This must have been due to this principle. Back then, hotel bathrooms in the States had infrared bulbs for heating in addition to incandescent bulbs for lighting. When they were on, my shadow would appear blue, and I had been wondering why that was.

I experimented in the bathroom and reached the conclusion that is was due to ocular deception that human eyes saw blue-green in the shadows as a complementary color to red.

After I had this experience, I was able to fully understand the dichroic method. Perhaps Dr.Land invented the dichroic method in the bathroom too.

The dichroic method caused a great flutter for a while, but I never heard about it being put into practice, so I suppose it was just something Dr.Land did for his own enjoyment.

Dr.Land became very good friends with Morita-san, and our technology exchange on instant cameras continued for many years. I too contributed a lot in solving problems such as the application of tablet batteries and long-range measurements using ultrasonic waves.

NEW TECHNOLOGY MUST BE DOCUMENTED AND RECORDED RIGHT AWAY

A few years ago, SONY was accused of interfering with a patent that Kodak owned in Japan, the States and Europe regarding the gap length of the video head. SONY was considered one of Kodak's targets when it purchased the patent that Dr.James Lemke had filed for with the intention of demanding royalties from all other video companies. If they had to pay royalties, all Japanese video manufacturers would suffer heavily. I read the details of the patent claim and was astonished that it was full of such common knowledge technology-wise, and was surprised that the patent was even approved. Anyhow, I had to search for documents to prove that it was public knowledge and filed a request for trial of invalidation.

Fortunately, my own article on "high-density magnetic disc cards" was published in the March 1975 issue of "Electronics" magazine in which I explained about video head gap lengths being used for recording at high density, and I was able to invalidate the Lemke patent using this official publication. What a relief!

In order to prevent such problems, people working in engineering should file their leading-edge technology ideas and submit them to the patent agency as soon as possible. Even if theydon' t get approved, try to publish them in academic journals and conferences to document it.

MORITA-SAN'S DETERMINATION TO BECOME "TOTSUKO OF THE WORLD"

Philips, based in the Netherlands, is the largest home electronics manufacturer in Europe. They started with incandescent lamps and moved on to fluorescent lamps, radios and record players, and then to tape recorders, contributing to the dissemination of cassette recorders while always leading the industry with their technology. When it came to color TV, their quality, design, costs and even production method were far more superior to that of their competitors, so they always stayed on top. Regarding the production technology of color TV, they constantly introduced new assembly methods to reduce costs and released products of high service performance and stability.

When Morita-san visited their headquarters in Eindhoven, Netherlands in 1953, he was encouraged to learn that even an agricultural country like the Netherlands was able to create an electronics manufacturer that has become known far and wide like Philips. He decided to make Totsuko not just Totsuko of Japan, but Totsuko of the world, and in order to do so, the company would need to focus all efforts on exports.

The Netherlands is not just an agricultural country, it is also home to a mega oil company called Dutch Shell. It is a beautiful country with a great deal of vitality, and what I am most envious of is that they have ten times more flatland than Japan per capita. Surely, they have one mountain that is about 200 meters in height, but other than that, the idyllic scenery sprawls endlessly. Since the surface area of the land is almost equal to that of Japan, and the population of the Netherlands is a tenth of Japan, it means that they have ten times more flatland per person. What a luxury! Even then, they were working on a long-term plan to expand their national territory by draining the lowlands in the Rhine Delta, and I was impressed that they were continuing the construction of the embankment.

SONY started a technology exchange with Philips in around 1968 when the CV-type VTR was released to the world, and I was working on my next project, the cassette VTR. Since then, I visited their offices in Eindhoven and Vienna where they have their VTR plant almost every day and discussed VTR systems, tapes and heads.

I recommended that they integrate their system with SONY's VTR system, but we did not reach an agreement. They were producing an open-reel half-inch VTR called LDL-1000, so they could not change their method at the time. We held discussions the following year regarding the development of the next version of our cassettes, but Philips wanted to use the tapes in halves, as side-A and side-B, and read/write both on both like what had been done with audio cassettes. Their concept did not match with ours, so we did not reach an agreement there either. Later on, a VTR based on this system was released as the VR-2000.

Philips aggressively developed many other new products and brought us the laser video disc in 1972 and a prototype of what would later become the compact disc in 1978.

LUCK DOESN' T ALWAYS COME AROUND TWICE

In 1980, a meeting was held on the standardization of SONY's video movie (8mm video). The following year, another meeting was held between five companies (SONY, Philips, Matsushita, Victor Japan and Hitachi), and a final agreement was reached on video movies. This was when I had the occasion to meet Mr.Lou Ottens, the development VP of Philips, and showed him our magnetic camera "Mavica." He said that Philips was researching CCD cameras as well but wondered "who would buy such an expensive camera using a CCD imager? "Mr.Ottens was in charge of VTR when I first met him in 1968, and he was also plant manager of the tape recorder plant in Belgium, inventor of the tape cassette and development leader of VTR, so he insisted on using side-A and side-B for the VTR cassette too. They released this VTR VR-2000 in Europe, but after various problems arose, they had to discontinue normal sales and sell VHS VTR by OEM. Luck doesn' t always come around twice.

As for the CCD imager, I don' t know if they failed to foresee that technology will progress, or if they thought that there was just no way Philips could make it happen. It seems that they had overlooked the luck that came around twice in this case.

Philips is a company with a very high degree of technical capabilities, but with VTR, it seems that it was not enough. VTR manufacturers disappeared one by one from Europe, and oddly enough, it was just Japanese manufacturers that remained in the VTR sector.

DEVELOPING THE "BETAMAX"

In 1971, the U-matic was released both in Japan and overseas at the end of March, and with it, I managed to clear two hurtles, the VTR cassette and color.

With no time to rest, I was assigned to concurrently work as the 2nd Development Division, which was quite unexpected. Iwama-san had been in charge of this department, and I was to take over his position entirely.

There were four sections in this department. The technology section was led by General Section Manager Kenji Yamagata and Technical Manager Kazuo Ehara. The productions section was led by General Section Manager Hiroshi Ohta and Camera Manager Takao Aoki, the procurement section was led by Division Chief Hiroshi Matsuzaka and Procurement Manager Hiroshi Shibusawa, and the engineering section was led by Manager Masashi Yoshida.

Morita-san test-driving the fuel battery bicycle

The department produced PV-type, CV-type, EV-type and DV-type VTRs, as well as PVC-type, VCK-type, IIC-type and DXC-type vidi-cons, image orthicon cameras, and various television monitors.

I had Yamagata-san, Aoki-san and Shibusawa-san, experts who had been in productions for a long time, teach me the basic rules of the production team and what they had been discussing in the meetings. But come September, just when I had barely started to understand what was going on in the department, I was released from my duties there. I was only there for a short while, but I am grateful to all of the people who taught me how tough production is.

The development themes for the 2nd Development Division in 1971 were quite varied, including things like bicycles that ran on fuel batteries, which the Research Laboratory had been working on. It is hard to imagine that SONY would have been developing something like this at one point, considering how the company is now. A prototype automobile using fuel batteries, co-developed by SONY, Shinko Electric and Fuji Heavy Industries, was announced in October, and we prototyped a bicycle as a test application of the battery. Unfortunately, we were unable to solve one of the major problems we had, which was where to set up battery fuel recharging stations for cars. So, with no chance of the fuel batteries going into production, development was terminated.

We also worked on the Sony Sports Clinic System, a device to provide integrated diagnostics of various sports activities, which was completed and released on October 27th at the Keidanren Kaikan. It was a system that combined newly developed slow-motion video and other slow-motion video cameras, as well as video recording/editing devices. Featured scenes of arms and legs joints moving were extracted from the video and the action was displayed in a continuous curve, allowing for better analysis of movements.

I was extremely busy at around this time, and since I had an equation for the Helical VTR tape format, I asked Yoshimi Watanabe to do all of the calculations for the 400-type format, and had him determine the dimension accuracy of the details. Based on this, I was able to design the mechanical section of the 400-type VTR. After several minor improvements, in November, I was confident that it was ready for production.

I showed Ibuka-san the cassette and said,

"Isn' t this just the right size for home-use? I made it the size of a paperback." To which Ibuka-san replied, "What I wanted was something the size of SONY's planner."

So I placed a cassette on top of a SONY planner and, finally, he was satisfied. Maybe it looked bigger that it actually was, since the thickness was nearly double of the planner.

I had completed the master format of the home-use VTR that I had always wanted to make. I had written in my 1971 SONY planner in November, "I want to mass-produce the 400-type," so I really must have been into this development.

I started working on the 400-type prototype the following year, 1972. I was always busy flying in and out of Japan gathering technical information. In May, alumite magnetic bodies caused a buzz. In June, an engineer visited us from RCA and we discussed the compatibility of the 3/4-inch VTR. In November, we attended a briefing at RCA in New York on the static video disc and a briefing at Philips on the laser video disc. It was a year in which I felt in my bones that the world was becoming more turbulent due to the development of new technology.

Since there was no way we could have fallen behind with development, we held a meeting on November 30th to determine the strategic direction for the 400-type. On November 12th, I ran a demo of the 400-type at the management meeting to help facilitate them making a final decision.

As a result, the leaders for the development project were determined in January 1973.

- Tape team: Manager Higuchi, Manager Iino, Assistant Manager Hisagen, Assistant Manager Ishida

- Head team: Manager Okumura, Assistant Manager Miyairi

- Cassette team: Manager Esashi, Manager Yoshida, Assistant Manager Watanabe, Kawamata

- Parts team: Manager Saito, Nishioka

- Interface team: Manager Tsuda, Assistant Manager Kawamura

- QA team: Manager Nukada

These members agreed to handle all issues and improvements and complete the project within a year.

On January 26th, we had a meeting with TEAC VIDEO, and having completed the U-matic portable prototype, we were to discuss the secondary prototype. TEAC VIDEO is a company that was established in 1971 as a joint venture with SONY and TEAC, and I, as the company's director, offered advice on development. In the meanwhile, I was also spending a lot of time on the development of U-matic machine applied equipment.

In addition, I was also developing the Filmpatrone system video machine. I completed development for that by April. It was a device that would record 50 images for 30 seconds at a time, displaying in slow-motion and as still images, both in color. Development for machines using this method continued for about a year after that, and we considered its possible usage for educational purposes, but since the educational video machine that we had been developing jointly with IBM was terminated due to the fact that it violated the antitrust law, the machines that used the Filmpatrone system were terminated as well.

During the period of February to August of 1973, I worked on the following projects.

- Jointly developed a system for pollution-free vehicles with Honda Motor at the Atsugi plant

- Ran a video disc experiment and completed the mother disc using a 6 micron cutter

- Produced the Sports Clinic System, ITOCHU to sell

- VTR 400-type image display completed in mid-March, started IC evaluation experiment

- Developed U-matic PAL and SECAM system

- Completed 5 units of sheet-method VTR for evaluation

- Completed TED-system video disc read / write experiment on August 23rd

"A CLUB WITH MORE CARRY"

In October, all of the Development 2 team was moved to building no. 200 at headquarters, which improved our development efficiency, and also improved confidentiality. There was even an exclusive guard checking staff member entry. In addition, we also created a room for doing research on golf, which Ibuka-san had requested a long time ago.

In this room dedicated to measuring golf-swings, there is a VTR capable of simultaneous image filming using four video cameras, one in front, one in back, and one to the left and to the right each, as well as devices to measure club-head speed and shifts in body-weight shift and a sheet recorder that displays slow and still images of the user's movements.

Ibuka-san practicing with the Golf Clinic System

Yoichi Odagiri and Tsuyoshi Hori took care of producing and setting up this equipment for the golf clinic room.

Famous professional golfers from Japan and overseas came to visit this golf clinic room in an endless stream. For example Arnold Palmer, Jerry Pate, Isao Aoki, Hisako Higuchi and Kikuo Arai all stopped by at one point or another. They enthusiastically did test-swings and discussed the problems with their swing forms. Most golfers were accompanied by Ibuka-san, and they enjoyed their golf-related discussions. Everything they talked about, from forms, timing, club grips to even about the mechanism of the club itself, was taken in by Ibuka-san in his own way. He would try to experience them on his own until he thoroughly understood these points, and if he was dissatisfied, he would come to use and give us problems to solve as an assignment.

When he asked us to improve some golf clubs so that the ball would fly farther when hit, this task was handled by the team to which Sohei Masuda belonged. Masuda-kun was from the Atsugi plant and happened to be developing battery-run bicycles in the Development Section 2. Since we needed a hitting machine, we asked Bridgestone to let us see the hitting machine they had imported from the States, and we constructed a machine based on that. We also asked Greenway Golf, a company located in Shinagawa with whom Ibuka-san had been on good terms, to create prototype golf clubs.

The barycentric position and initial velocity of the sample clubs with different loft angles of the club faces could be measured in the measurement room, but the course and the distance of the ball had to be measured in the actual links. So, we borrowed an open space in the Kota Plant in Aichi Prefecture to hit golf balls and obtain data.

TECHNICAL PREPARATORY OFFICE ESTABLISHED

On November 13th, 1973, a strategic meeting on SLX was held, and the project 400-type was to be SLX, with a group to be established to consider full-scale production. With Fumio Kouno (now senior managing director) as the leader, and talented engineers who had been transferred from the Development Section 2 including Serizawa-kun as the core, the Technical Preparatory Office was launched.

In 1974, the Mavica-system that we had been prototyping for about six months then had been improved enough to be used as an educational tool, so Iwama-san ordered us to start full-scale development. In addition, we also needed to start developing the talking card (magnetic recording card), as Ibuka-san said he wanted to make it widely available around the world, especially for illiterate people. Therefore, I was to work on the development of an educational tool using the talking card in addition to the development of the 400-type.

On May 8th, we announced that we had developed a video player that can easily read/ write color images using a magnetic card.

On May 14th, we presented a paper on Mavica at the IEEE International Magnetics Conference held in Toronto, Canada.

In June, Ibuka-san (chairman) announced his perspective on the talking card.

- (for Iran) More than a few million cards should be sold. We would like to keep the price to about 30 dollars.

- (for Japan) An educational tool based on the card should be developed. It should be usable in language labs for adults.

- (others) Build up track records starting from Iran, and appeal to UNESCO. By July, there should be at least ten units of the T-card player

On July 24th, I accompanied Ibuka-san to display educational equipment at the UNESCO Headquarters in Paris and ran a demo of the talking card. The 1974 Electronics Show was held in Harumi in SeptemberThere, we displayed video equipment, including Mavica. His Imperial Highness visited us, and I explained the technology to him, after which he commented that "the videos have become much smaller now."

Starting on November 6th, I accompanied Ibuka-san on a visit to visit the Philippines, where we showed president Ferdinand Edralin Marcos the talking card and other educational tools. As a result of this visit to the Philippines, it was decided that video software was to be distributed as an educational tool in the backcountry, and several years later, the Betamax was to be used in bulks. Initially, they decided to distribute instructional videos on how to use contraceptives to prevent population increase.

"VTR PV-100" TO BE IN PERMANENT REPOSITORY IN THE SMITHSONIAN MUSEUM

In October, SONY donated 16 breakthrough electronic products, including the Esaki diode, the world's first all-transistor VTR, the PV-100, the world's first all-transistor B/W television, and the Trinitron color television. These were chosen to be placed into permanent repository in the Museum of History and Technology. I consider this to be a great honor.

In January 1975, an electric wheelchair was built with Masuda-kun of Development 2 as the leader for the "Japan Sun Industries," a house for the physically disabled located in Oita, Kyushu. This wheelchair was designed so that it could be assembled at the house, so this technology was transferred to Japan Sun Industries right away, including jigs and special tools developed so that it could be assembled by people with disabilities, and the production line was launched smoothly.

On June 2nd, we received His Imperial Highness at the Japan Sun Industries and showed him the production line of the wheelchair being run by the disabled, the actual usage of the talking card at the education corner and the rehabilitation at the sports clinic corner.

RELEASING THE "BETAMAX"

About a year after the technical preparatory office was launched in 1974, the extraordinary efforts of Kouno-kun and the engineering team bore fruit, and the secret code of SLX was determined to be "Betamax VTR" and to be disclosed on April 16th, 1975. The final decision on whether to use the H-skip color recording method (skipping color by each scanning line) or the color frequency interleavemethod (low-pass phase) derived by Amari-kun right before the commercialization was made by Kouno-kun, who decided to go with Amari-kun's method. These problems were resolved without incident, and the day of the release soon arrived.

That day, we held a press release on the 10th floor of the Keidanren Kaikan for the journalists at the Keidanren at noon, journalists of industry journals and foreign news agencies at two, and software makers at four, explaining the features of the Betamax and demonstrating the device.

With the release of the Betamax, the industry's reaction to seeing the start of the home-use VTR era was intense. The Nikkei Sangyo Shimbun reported as follows:

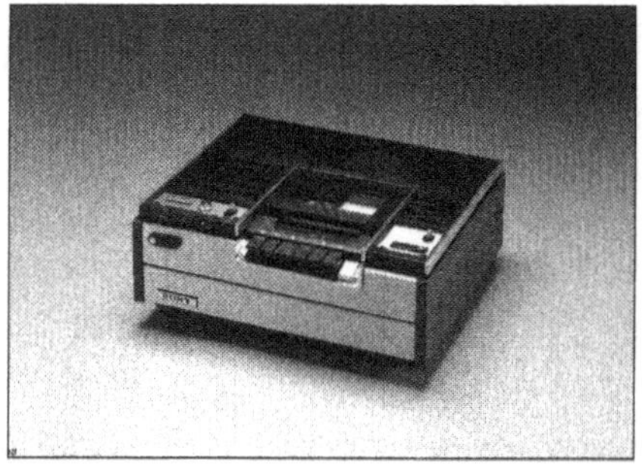

Betamax "SL-6300"

"Already more than a decade since video was said to be the likely winner of post-color - industry participants had all been making desperate efforts to grasp the demand at home, which is incomparable to the industrial market, but they had not been able to overcome the issue of price. However, with the new product development, SONY suddenly succeeded in lowering the price by more than 100,000 yen and is now poised to go on the offensive in the home-use market. On the other hand, competitors who fell far behind SONY will suffer greatly. Some manufacturers have already started a half-price sale as if they are reducing stocking, but whether they can catch up technology-wise to SONY any time soon or not is a difficult question."

The Dempa Shimbun Daily reported:

"The biggest feature of the Betamax is the price, which is by far cheaper than conventional cartridge and cassette VTRs. It gives the impression that it has become much cheaper, and it has become much easier for the dealers to handle compared to conventional products. Needless to say, SONY plans to offer a wide variety of its Betamax series, including a portable type, built-in tuner type and play-only type. The company has also clearly expressed its course change from their previous policy to stick to the U-matic method to find opportunities in both industrial and consumer markets, to a policy to use Betamax for the home and U-matic for businesses."

So in May, the Betamax went on sale, and the price was 449,800 yen for the Betamax system combining the SL-6300 video deck and an 18-inch Trinitron, 229,800 yen for just the video deck.

THE "SONY IDEA CONCOUR"

When the Betamax went on sale on May 10th, 1975, I accompanied Ibuka-san to attend the Fifth All-Honda Idea Contest hosted by Honda Motor's Suzuka Plant.

This idea contest had been held since 1970 as an occasion for all Honda employees to present their originality and ingenuity. It has grown year by year since then, and in 1975, there were 5,000 entries from which 51 ideas were selected as finalists. Ibuka-san and I were invited to offer our opinions and suggestions on these ideas and the final selections.

The contest was held from 9:30 to 16:00. All entries were devices that used either manpower or engines, and some were able to make movements that were comical, strange, impressive, or just plain practical, all of which were fun to watch and fun to ride. It was so much fun that time seemed to just fly by. Honda-san and Ibuka-san also seemed to be enjoying the event very much, test-driving the various devices.

"Mr.Eiichiro Honda test-driving one of the ideas
at the All-Honda Idea Contest"

Of all of the entries present that year, I gave top marks to a device that moved like a centipede, while I found a bicycle with semi-circular tires the most amusing among the unusual bicycles.

Inspired by this Honda Idea Contest, SONY also organized an all-company event called the "SONY Idea Concour," the first of which was held on September 14th at the Atsugi Plant.

As one of the judges, I evaluated all of the creative works filled with a playful spirit, and it was very difficult to give them points since all of the entries were the results of great effort. But the one that stood out the most was a manpowered helicopter. It was Yuji Nagano Manager, my subordinate at Development 2, who invented and created it. Nagano-kun was a mechanic who designed the mechanical section of the TC-201 series, a product that was said to be one of the best of the open-reel tape recorder era. I was amazed to know that what he poured his passion into making was not a small tape recorder, but a huge helicopter.

Manpowered helicopter which received an award
at the SONY Idea Concour

This helicopter did not fly, but could only hover for one second about 20cm off of the ground, so it was nothing close to a world record, but the operating principle was unique, which won him the highest points.

The propellers were turned by manpower, until the flywheel effect pooled enough energy into the propellers, after which the hook on the central rotating shaft released the propellers. The propellers flew outward due to the centrifugal force, but there was a cam attached which changed the pitch of the propellers. The propellers first pooled energy through the flywheel effect, and then after the hook is released, automatically changed the pitch and jumped up in the air.

A light-weighed sprinter was chosen as the pilot, Seisuke Suzuki from Development, and he experimented repeatedly in the open space and practiced until he learned the ropes before he successfully flew 20cm high in the gym hall of the Atsugi Plant.

VISITING THE NATIONAL DEFENSE FACILITIES IN THE UNITED STATES

In November, I was chosen as one of the 20 members of JDS to visit the national defense facilities and strongpoints in the United States from the 18th for about two weeks.

JDS stands for Japan Defense Society, and numerous companies, including SONY, were registered as members. This visit was an invitation from the American war veteran association, and we were able to observe closely the reality of the country's national defense.

On the 20th, we visited the Vandenberg Strategic Air Force Space Command and saw the Titan II rocket launch site, the launch control room, the experiment observation room and the rocket assembly site.

We were told that the Vandenberg Air Base tested rockets a few times every month, firing mainly in the direction of Hawaii. They usually just did tests here, but they said that in case of an emergency, they had everything they need to actually execute missile attacks. But the main missile bases were located in the northern mainland and in Alaska, and that reinforcements were being done.

There were six missile launch sites in the bush, and we could see a hole of about 8m in diameter and 3m in depth. An iron plate about 50cm thick would slide sideways to cover the hole.

About 50m from that launch site was the entrance to the launch control room, and when we went down to the third basement floor in the elevator, there was a small room with two launch buttons. One officer and two soldiers stood on call in this room for eight hours, so that whenever the president gives the order in case of an emergency, missiles can be launched immediately.

Seeing this, I felt sorry for these soldiers who stood all day in front of the buttons in a deep dark hole like this for some occasion that may not happen in the next ten years.

But upon hearing the next explanation, I was shocked. The two buttons were set about 3m apart, and could be launched by two people pressing the button at the same time on the count of three. If it was just one button, there could be a risk of one person pressing it on accident, or them being a spy or having gone insane, but if the missile is launched only when two people pressed the button at the same time, the risk of accidental launch is almost zero.

What if the two soldiers were both spies or became mentally unstable at the same time? In that case, the rulebook stipulates that the officer watching from behind should shoot any suspicious soldiers dead.

I became painfully aware that their strict shift had been put in place not in times of war, but in our current time of peace.

On the 21st, we flew from Denver to Colorado Springs and headed for the rocky Cheyenne Mountains by bus.

There was a rigorous security check on the way, but we were allowed through since we were members of the JDS and arrived at the foot of the mountains. There, I was astonished to see a huge tunnel with a two-lane road going through it. I had been told about it before we arrived, but I had never imagined a road like a highway could be built inside of a mountain. This was the entrance to NORAD, the North American Aerospace Defense Command.

About half a mile into the tunnel, there was a huge door which majestically opened, to a three-story building that looked like dice came into view. I looked down and was surprised to see that the building itself was floating off of the ground on huge coil springs. A steel coil that was about 10cm thick was wound into a diameter of about 3m, and was lifting the building at least 2m off of the ground. I couldn' t tell how many springs there were because it was dark in the back of the tunnel, but they seemed to be innumerable. Everything was lighted, including the building, so they rose up clearly in the dark tunnel.

Entrance to NORAD, 1975

Next, we took an enjoyable flight from San Francisco to Honolulu to visit Hawaii.

The flight was on United Airlines, which had been hosting a quiz show, with the first prize being champagne and an airline bag.

I didn' t know that there even was such a quiz game, so I read the pamphlet that had been set on my seat, which said, "please guess the exact time when this plane passes the halfway point from its departure to its arrival. Those who guess correctly will win a prize." I love these types of intellectual quizzes, so I looked at my watch and listened to the in-flight announcements to note the time of departure, the distance to Hawaii, the flight speed, the jet stream speed being tailwind and so forth. A while later while eating lunch, I realized that there were two catches in this quiz.

About two hours before arrival, the answer sheets were collected, and the pilot announced the answer and the winner. It was my name that was called.

A cute cabin attendant came over with champagne and a cylindrical-shaped United bag in her hands and gave them to me, saying, "congratulations." I felt elated.

But the other members of the JDS watched vacantly, not understanding what was going on. Eventually, they found out about the quiz and said, "oh, I didn' t know there was such a quiz being held. Lucky of you to have won the prize," they said regretfully.

Perhaps because of their age or because they were tired of work, their instincts did not work very well for games like this. But even if they had noticed that there was a quiz, and even if they could understand the English announcements, they wouldn' t have been able to answer correctly unlessthey knew the time difference between San Francisco and Honolulu and Pacific and the Hawaiian standard times.

These two factors were the catches that prevented people from answering the quiz easily.

There was no way that the States would have just shown their top secrets to us Japanese, so maybe they wanted to show off their military strength to discourage our desire for war. For me, it was a very interesting experience when I think about this in connection with the industrial war.

"BETA FORMAT" STANDARDIZATION ENDING IN FAILURE

The year 1975 ended with the release of a paper titled "Development of a New System of Cassette type Consumer VTR" written jointly by me, Kouno-kun and Yoshio Ishigaki (now president of AIWA), which we presented at the IEEE Fall Conference held in Chicago due to the global focus on the release of Betamax.

This paper received the Best Paper Award at the 1977 Chicago IEEE Spring Conference.

1975 was a year in which many unforgettable events came up one after another, and I had no time to rest, always working on development, preparation and presentations.

I had been attending technical meetings when necessary as a member of the development personnel of the Betamax, but unfortunately, I was never informed about top-level meetings and their decisions on standardization.

The situation at the time is explained in a book by Mr.Yasuzo Nakagawa titled "The Development of Magnetic Recording in Japan," written based on research from an unbiased standpoint, which I will cite here.

It is from a section about SONY calling for the standardization of the "Beta standard."

It was the fall of 1974 that SONY's Kihara created the "Betamax," a method which arguably was the ultimate solution to home-use color VTR.

Along with this VTR, a technology to incorporate color signals using the azimuth method, which was thought to be impossible until now, was created. As a result, the surface recording density dropped drastically to two square meters per hour, enabling downsizing and weight trimming. This was a breakthrough event in the development of magnetic recording technology.

Before the release of this VTR, SONY offered joint development to Matsushita and Victor Japan, both of which SONY had been cooperating with on the "U-standard" for professional-use videos, showing them the blueprint. In addition, by early December, top management of the two companies were invited to SONY headquarters in Kita-Shinagawa and shown the actual Betamax. The aim there was to standardize home-use VTR to maintain compatibility, which SONY considered a priority. That itself was a meaningful gesture. However, the problem was the attitude of SONY's top management who met the guests. Shiraishi of Victor Japan reminisces,

"We vaguely knew that SONY was developing a new device for consumers since there had been rumors her and there. But until we were shown the actual product, we had no idea what it was really like. Seeing it with my own eyes, I was impressed and thought SONY was true to its reputation. That is how good the device was. But SONY said confidently, 'we made such an excellent product. Why don' t we make this the standard? ' Myself being an engineer, this sort of concerned me."

Sugaya of Matsushita, who accompanied his superiors and saw the product on a different occasion, also recalls, "SONY's explanations implied that their new device was the best thing ever." There might have been a difference in their company culture. At the same time, it is also true that the psychology of the passive side had a great effect. In other words, that is how confident the leaders of SONY were about the quality of the Betamax. It is ironic that their confidence in turn appeared coercive to others."

(Yasuzo Nakagawa, "The Development of Magnetic Recording in Japan," Diamond, Inc.)

In my view, SONY was showing the Betamax to other companies from an early stage and answer technical questions so as to somehow make everyone agree to the standardization of the format, and I think that they met their guests in a sincere manner.

Both Morita-san and Iwama-san contacted the leaders of Matsushita and Victor Japan repeatedly to ask about their response on the standardization of the format, but after more than six month, there was no response at all.

Morita-san must have been tired of waiting. He finally made the decision to put the Betamax on the market in May 1975.

Even then, SONY continued to try to hold discussions on the standardization of the format without giving up. I have picked out memos regarding format standardization from my 1976 notebook.

- February, SONY sent an inquiry to Matsushita regarding standardization, to which VP Nakao responded that they needed another three months before answering.

- March, Mr.Konosuke Matsushita contacted Chairman Morita to take a look at Victor Japan's VTR, and I accompanied the chairman to Osaka to meet Mr.Matsushita (advisor). He then told us his final decision, "I have compared and considered your Betamax and this VHS, and since VHS appears to be cheaper to produce, we have decided to go with VHS."

- April, another meeting on standardization was held at Victor Japan headquarters in Nihonbashi, Tokyo, and Chairman Morita, President Iwama, VP Ohga, Kouno and I attended. However, there was no compromise made in that meeting, and the discussion ended in a rupture. We had brought a Beta machine capable of two hours of recording to demonstrate that it was possible to produce a two-hour recording device, but their determination was solid, and our effort were ultimately in vain.

This was the 5th of April. But then in my notebook, it says that on the 9th, "Victor visits SONY to see the Beta portable on the 9th floor showroom." I don' t remember clearly why they visited us, but I imagine that this was the day that the engineers exchanged opinions for the first time.

I was only told later that President Iwama revisited President Matsuno at Victor Japan headquarters to make a request for standardization, but the discussions broke down again.

THE GLORIOUS "30TH YEAR SINCE THE FOUNDING OF SONY"

The year1976 marked the 30th year since the start of Totsuko, which changed its name to SONY.

On this occasion, a new administration was announced for management and the personnel system was structured as follows:

Masaru Ibuka (Chairman) to be Honorary Chairman

Akio Morita (President) to be Chairman

Kazuo Iwama (Vice President) to be President

Norio Ohga (Executive Director) to be Vice President.

At the board meeting in April, the "Ibuka Award" was created to be awarded each year to contributors to SONY's technical development, and I was chosen to be the honorable recipient of the first Ibuka Award. The reason for my selection was the overall development of magnetic recording.

Prior to this, I had received a special award with Yasuda-san at the Okuma Auditorium of Waseda University on SONY's 15th anniversary for the development of tape recorders.

Later, I was also awarded the fourth Ibuka Award for the commercialization of the overall Beta system, along with Keizaburo Tozawa. I am very grateful for this.

CHAPTER FIVE

HOW I THINK
The Philosophy that Engineers Should Have to "Fulfill Their Dreams"

1. IT IS NOT AN ERA OF SOLO DEVELOPMENT

MAKE IT HALF OF BETA THE NEXT TIME

The year 1976 was the 30th anniversary commemorative for SONY, and I was to announce the development goals for that year at the research and development briefing session held on January 19th. There, I submitted four new development topics.

(1) PWM (pulse width modulation) power-amp

(2) Jacket Mavica

(3) CCIR (for the Europe model) Betamax

(4) The next-generation of video after Betamax

Among these, the next-generation of video after Betamax was a subject natural for me.

I had a way of working that I always followed ever since my days of tape recorder development. Whenever a new product was developed and handed over to production to take to the market, the development team should immediately find its next development product, and create next-generation products with features that previous models did not have.

When the open-reel CV-2000 was acknowledged internationally as a new product, I handed over its development to the production team right away, and everybody in video development turned their focus on developing cassette video, which is how we completed the

U-matic.

Although video was in cassette and color now, it was still too expensive and too big for home-use, so it was only used for industrial purposes. It was November 1971 that the concept of the home-use 400-type was put together and development started, just around when the U-matic started to become popular in the world. It was only after three and a half years in May 1976, when the U-matic became SLX, then released as Betamax, that the home-use video was born.

DEVELOPMENT RACE BETWEEN TWO TEAMS

If we were going to plan the development of next-generation video right after Betamax was released, we needed to have some sort of meaningful goal.

In this case, our number one goal was to make it portable. While I was working on the development of videos, I started thinking that the silver-salt photography method will gradually shift to the magnetic method. It was not that movies and photographs would disappear, but that, in this era of television, it was clear that a handier form of video would be what penetrates the home market, and when a small and portable video player appeared on the market, 8mm films would be eclipsed.

The 16mm Eyemo Camera that had been known as a professional movie camera was rarely used after the U-matic ENG appeared.

A while before this, filming with 8 mm was the trend, and since my hobby was filming with an 8-mm camera,

I would go hiking and film things in the mountains a lot. Then, I started enjoying making 8-mm talkie films by applying magnetic coating to the 8mm film. Then, my hobby developed into a full-fledged invention that mechanically connected a tape recorder and an 8mm cinematograph and synchronized the image and the sound to make it a talkie. This was eventually released by SONY as a product called Film-Tape Synchronizer (FTS).

How time flies. The 8mm now seemed so old, but I came up with the idea of changing it into high-definition, high-quality sound portable video.

In fact, I had always thought that the ultimate form of video was the size of 8mm and its portability, so if I could make this dream come true, this would make for the ultimate video, and thus was an idea that I must succeed in developing.

So, I decided that the ultimate form of video would be half the size of Betamax. And thinking that it would save a lot of trouble if I just cut all of the numbers in the dimensions of the Beta mechanism blueprint in half, I decided to go with that idea. If the dimensions were cut by half, the volume and the weight will be one-eighth, making it a "portable" device. And if the design was original, I was confident that the impression of video would change.

Thinking that the development team should put all of its efforts into the development of this ultimate form of video, and expecting it to be a long-term project, I thought that it might be a good idea to create two groups and make them compete on the same project in order to educate the engineers. I discussed this with the leaders of Team 1 and Team 2 and had this tactic further executed separately within each team.

Actually, I began thinking about training my subordinates at around this time. Until then, I had always worked on my own, and I had conceived, designed and tested new products by myself. Then, I handed ideas that seemed to work to my subordinates to continue working with the production design.

This was how I worked up to this point, but when I thought about it, I realized that engineers cannot learn to conceive new ideas and develop products this way. I had always been too busy—rushed to develop products, always pressed for time—but for this video development project, a several-year cycle would be necessary.

My spearheading things all the time was doing little good, so I gave assignments to the two teams so that as many engineers as possible could compete with each other while contributing ideas, brushing up their skills and learning the know-hows of engineering.

I wanted the engineers to learn about my unique way of conceiving ideas from my long history as a developer and utilize this in their work, so I made a speech entitled "The Secret of My Ideas" at the memorial lectures held at the "Sony Idea Contest" on February 19th 1976.

The topics of the speech were as follows:

(1) Hoist your sails when the wind is fair,

(2) gestate your ideas well,

(3) set a target,

(4) colds can be useful depending on how you suffer them,

(5) pain is the source of pleasure,

(6) be of good birth,

(7) never copy others.

I will talk about these later, but before that, I would like to talk about the two competing teams. About six months after starting my plan in August, the leader of Team 2 came over to me and said, "We couldn't downsize it to half, so we made a 70% sized prototype." He brought me the device and replayed an image. Looking at it, I immediately said, "How can you say you made it smaller at just 70%? If it's not 50%, I'd be too embarrassed to call it a next-generation device. This is no good. You should do better."

I asked Minoru Morio (now Executive Deputy President) of Team 1 about his opinions, and he said, "Please give us the time to make it half the size." Of course, I knew six months wasn't enough to make it happen, and there is no easy way to do it other than trying every method conceivable and building up one's skills, so I told him that we should work on it at all costs no matter how many years it may take.

From then on, they worked very hard motivated by my pet phrase "have an at-all-costs spirit," until the product was released in 1980 as the "Video Movie." In the four years after I started the halving project, we maintained close communication with the tape development team and the head development team in the Sendai Plant, and, finally, we succeeded in halving the head gap length and the shortest recording wavelength.

We were able to use both metal particle tape (MP tape) and metal evaporate tape (ME tape), enabling high density recording according to usage. For the head, an amorphous (non-crystalline) head and the Sendust (aluminum and silicon alloy) sputter head were developed, resulting in a high output.

In addition, the color CCD camera, which is essential for video movies, came into use at the perfect time. We were very lucky with this.

"CCD IMAGING DEVICE" DEVELOPMENT

Color cameras were essential for home-use video, but since the cameras then used vacuum-tube image tubes such as Vidicon and Plumbicon, so they were far from compact cameras. SONY created a double-lumen full-auto compact color camera, the DXC-5000, in 1969 and employed it in the U-matic VTR. Later, Betamax videos used the home-use color video camera HVC-80 which employed Trinicon image tubes, but vacuum-tubes just did not fit the video movie that I was aiming for.

Celebrating the release of the new CCD product
with President Iwama

But there was one predecessor who tried to make his own dream semiconductor imaging device in SONY. It was Mr.Shigeyuki Ochi, now board director of SONY and director of Media Processing Laboratories. And it was Iwama-san (then President) who oversaw Ochi-san's research results and showed such enthusiasm that he made a bet-the-company decision that "this is the project that SONY's semiconductor division should work to complete."

This CCD (Charge Coupled Device) imaging device had been researched in labs all over the world, but it was thought to be almost impossible to create more than 100,000 imaging devices with one chip using large-scale integrating circuit technology.

However, Ochi-san's group of researchers overcame numerous barriers to create a one-chip 110,000 element CCD, which was a very large scale in March 1977, and announced that they succeeded in developing an individual color video camera. One article said that "SONY spent five years self-developing three CCD elements of red, blue and green, integrating approximately 110,000 semiconductors onto a 10. 3 x 9. 1mm silicon chip."

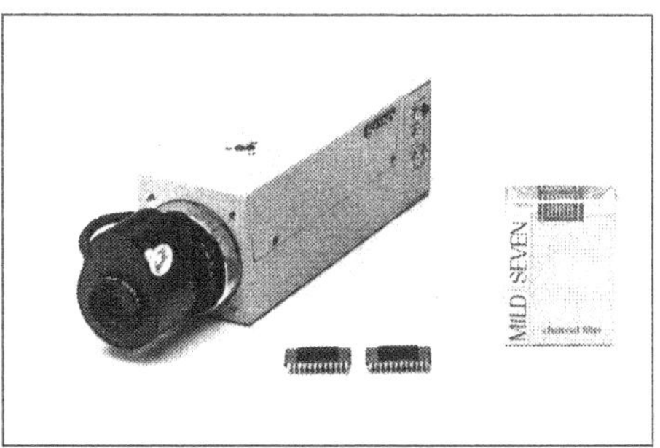

CCD color video camera "XC-1"

After that, two more years were spent studying how to increase the density, and by May 1980, a 280,000-element CCD color camera was produced. Based on this, a new color wide-projection (enabling the projection of a cinemascope-sized image) and the world's first in-flight image system using the CCD color video camera was born . It was installed in a Boeing 747 owned by All Nippon Airways where it received favorable reviews.

Thanks to these achievements, Ochi-san received the Invention Award from the Japan Institute of Invention and Innovation, in addition to Emmy Awards in 1990 and 1993.

RECOGNIZING THAT THE CCD IS "OF GOOD BIRTH"

Ever since SONY released its CCD imager, competitors seemed to have been frantically working on trying to develop semiconductor imagers, but all epigones other than CCD were inferior in performance. For example, the in-flight camera system installed in JAL's jumbo jets did not have a very good reputation, and it seems that it was not used in video cameras either.

Thus, it was due to President Iwama's insight in recognizing that the CCD imager was "of good birth" and his choosing from the beginning to continue with its research that the portable video era today first started.

In July 1980, two months after the news reported that the jumbo installed cameras had been released, SONY also released a prototype of the "Video Movie," a breakthrough camera/video system integrating the CCD color camera and an ultra-compact video cassette recorder.

The world's first 8 mm cassette recorder "Video Movie"

President Iwama suggested at this time that, "in order to commercialize the Video Movie, we first had to standardize the size, shape and recording method of the cassette, and that a standard format should be decided upon through discussions with other companies."

In order to execute this suggestion, we invited home electronics manufacturers, tape manufacturers and other associated parts manufacturers from all over the world to a round-table gathering about the 8mm video. We studied and discussed the finer points until everybody agreed on the technology and came up with some prototypes. As for the tape compatibility, SONY had produced a benchmark tape, with which they kindly provided us. It took more than four years to standardize the format, but finally in January 1985, the first 8 mm video, CCD-V8, was released.

1980, Mr.Michel Poniatowski,
Special Advisor to the President of France
(Ibuka-san loved to show newly developed
products to his guests)

As with the U-matic and the Betamax, it took a very long time to reach any agreements. This may have been because the times had changed. In the past, there was a phrase "founder's profit," which meant that whoever created a new product could enjoy its benefit. But doing so became more difficult over time, with margins becoming thinner as soon as competitors started selling competing products. Now, times were tough, and just like with the 8mm, everybody lined up at the starting line to start the race all at once.

ENGINEER SPIRIT THAT DOES NOT ALLOW FOR FAILURES

The release of the Video Movie was handled by President Iwama in Tokyo, and Chairman Morita at the Plaza Hotel in New York, with me acting as his assistant, both on July 1st.

I had always accompanied Morita-san on important press releases as an assistant so that I could help when the device needed to be demonstrated or if technical questions came up. I started preparing for the demonstration of the device at noon, did the press conference at 15:00 and filmed Morita-san giving explanations with the Video Movie or close-ups of the smiling faces of the female journalists in the front row, replaying these images immediately after Morita-san finished talking.

When the images were played on the screen there, everybody applauded, which surprised me very much despite my supposed familiarity with press releases. People were impressed that a small video camera could film images with much more detail than the 8 mm for such a long time.

The news of this spread and NBC News promptly requested that we run this topic in their evening news program on the 3rd.

Morita-san and I went over to the NBC studio at 17:30 that day. The playback monitor and the video player were arranged so that the performers and the video units would be in a semi-circle with the news caster in the center. I was told to hold the Video Movie to film Morita-san and operate the video tape while the reporter talked and play back the filmed image onto the monitor. Having finished the briefing and prepared everything beforehand, we waited for the actual show to start at 18:00.

About three minutes before our turn was to come, the producer suddenly came over to use and said he wanted to change the layout so that everybody would be closer to the center. A few young men ran over to forcefully rearrange our chairs and the monitor.

The show started, and I did the filming with the Video Movie. I set the tape in the player, rewound it and then pushed the play button, but the image did not appear on the screen. In the spur of the moment, I stopped the tape and did the following.

My brain was assessing the situation faster than a super-computer. The tech guys must have pulled the cord on the monitor, and the connection must have become loose somewhere. The power is on, the signal code is on, but the connector looked just a bit slanted." That must be it!", I thought. The connector was slightly loose. In a split second, I reached out and pushed the connector back in. Now the image should come.

With a cool expression, I pressed the play button again, and the image was replayed. All of this took place in the span of perhaps two seconds. I wonder if the viewers noticed my failure, or rather, my recovery technique. My hands were shown in a close-up, so people might have thought, "wow, what complicated actions this machine requires." Live broadcasts makes me nervous since there are no second takes. This must be why Morita-san makes me act as his assistant, since he has confidence in me. I was relieved that this was the most challenging thing that happened.

Back in my hotel room alone, I said aloud, "Thank heavens!" and I remember tears coming to my eyes.

FAILURE IS UNACCEPTABLE IN EXHIBITIONS

Most of the products that I developed were leading-edge products, so they were golden apples for the company, something that could not be damaged. When showing them to people in press releases and exhibitions in particular, they must be in their best and brightest condition.

When demonstrating our latest products to journalists and experts in the field, it is impermissible to make mistakes. If our top leaders introduce a product with full confidence and it doesn't work, the company will lose face.

Engineers in charge of displaying products for exhibitions must know this as a matter of course, but unless you have learned the true significance of this firsthand, you might experience some unexpected failure if you take your responsibility too lightly and assume that your product will work.

You should be able to anticipate all eventualities and have contingency plans in place in case something comes up. Differences in environment between the company laboratory or workshop and the exhibition site could also become a hidden pitfall.

For example, it is common sense to prepare a variable transformer (slidac) considering possible variations in power voltage at the site. The next major cause for malfunction is temperature change. Especially with new prototype ICs, devices may malfunction due to changes in heat dissipation conditions or the device heating up after leaving it on for a long time if you fail to check the temperature.

One thing you need to be careful of with the helical scan video is changes in temperature. In an exhibition hall, the humidity could increase due to the large number of people, and the tape might not be able to run due to vapor adhering to the friction surface of the tape and the rotation head. This is the same principle that works when you walk in from the cold outside into a warm room and your glasses fog up with water vapor. This makes the tape stick to the rotation drum.

Induction noise from the power source could also cause a malfunction when it is moved from one place to another, so we would always be extra careful and take measures. This happened one time at an exhibition in New York, which took place in the basement floor of the Americana Hotel. We set up the U-matic videos and Trinitron monitors and replayed images, but every now and then the images would shake wildly.

This was not suitable to show to our customers, so we divided into groups to investigate the cause of the problem. I was assigned to check the power fluctuation and the magnetic field fluctuation. While doing so, I noticed something odd.

Whenever the subway passed on the other side of the wall, the image would shake. The Trinitron set up near the wall on the side of the subway shook wildly, and the television set up near the opposite wall did not shake as much, so I suspected the subway was root cause.

Relocating our gear, however, would prove problematic. We moved the image replaying corner to the far side of the room, and the others to the subway side, and barely made it in time for the start of the exhibition.

With the U-matic, we had another phenomenon happen because of the cassette, and the designers were astonished that such a phenomenon had occurred.

This was after the explanation of the newly released product had ended, and the journalists started taking photographs of the product. There was one journalist who came up closer to the product to get a better shot and was using the flash to shoot from different angles. At one point, one of the devices that had been moving suddenly stopped. The U-matic used a transparent tape leader as the tape-end sensor, and when that sensor caught the light that came through the tape, the tape stopped. The strong strobe flash had passed through the spaces between the device, and the sensor caught the light and stopped the device.

It was caused by the strobe flash. We solved the problem on the spot, but we also learned an important experience from our customers, and we made improvements to upcoming products so that similar errors would not happen again.

After this incident, we developed a better sensor using metallic foil, which prevented from this type of problem from happening.

HOLDING A CONSECUTIVE NO-FAILURE TITLE (NOT TO BRAG)

From the Totsuko-era to today, there have been no less than 50 occasions on which we exhibited prototypes and new products to outsiders, but I do not remember failing in any one of them.

I don't mean to brag, but I think it was my engineer spirit that made this possible. I repeatedly ran all sorts of tests in order to prevent failures in the products that we put all of our efforts into, even using a thermostat bath to measure temperature attributes, changing the temperature to monitor changes, making sure that the devices do not malfunction whatsoever, preparing at least three backup machines just in case so they can be replaced if the event that something happens, and always erring on the side of caution.

I consider myself to be just another one of the engineers, so I could see clearly that my subordinate engineers thought the same way. They earnestly prepared exhibitions for the products that everyone worked so hard to create, and if a problem came up, they were willing to spend all night to solve that problem. Everyone prepared so as not to fail.

When I would give pep talks like "Let's make this presentation a success like always," they would say, "Oh, here comes the pressure again," but contrary to their whining tone, their faces were all smiling, as if they enjoyed the challenge.

BOSSES SHOULD UNDERSTAND THE MINDSET OF ENGINEERS

I was an engineer; never a boss. This is because younger people have newer skills and ideas that I don't have. Anyone who has greater skills than I is a teachers to me. In my opinion, a true boss must have excellent credentials and be able to win the trust and respect of his subordinates. Therefore, I am always learning new skills from many young teachers, forcing my old mind to study harder so that

I can be on top of the latest technology. But I am not making much progress, which is why I don't consider myself to be a boss.

So then, who can be called a boss? That would be people like Ibuka-san. Ibuka-san has a deep understanding of most people, but particularly of engineers.

I have never been forced by Ibuka-san to do something. Instead, he has given me many goals. An order is a one-sided instruction that is meant to be followed mindlessly, and if you don't, you will be devaluated or fired.

But a true engineer would gladly work on freely seeking out new ideas as long as they are given goals. A true engineer is always waiting for a goal, not an order.

Please do not forget that engineers devote all of their energy to achieve their set goals. Ibuka-san knew this very well and would heartily congratulate whatever the engineers made because he know they had done it with their heart and soul.

On the other hand, there are bosses who do not understand the mindset of engineers. In general, there are perfect bosses who have a great memory, vast knowledge and common sense. But the following examples may be useful to understand what the engineer mindset is like.

(1) Please do not do anything that might hurt an engineer's pride. They have gone through unspeakable efforts and hardships, thinking of everything conceivable in order to choose the best method to make a given product. Therefore, they have great pride in the products they make.

For example, one group of engineers created a new printer, and someone who saw it said, "This digital print is worse than photographs. If you can't make it better than photographs, why don't you just quit?" He might have been joking, but this hurt the engineers' pride badly.

I told these engineers to "continue with your research. Make something that's even better than photographs." They continued working on it for a year, and when the person said, "You must be kidding. Are you sure you didn't just glue a photograph on paper to show me?" I was relieved to think that the engineers were like elated on the inside.

(2) Please trust the engineers. The last thing a boss should do is to show off knowledge as if he knows everything. Rather, a sense of trust is deepened when the boss keeps his mouth shut, listening with a smile to whatever the subordinate is saying. Even if you do have the knowledge, you should not get ahead of your subordinate's explanation and blabber from the sidelines.

There was an old TV commercial in which a wife asks her husband, "Have you cleaned the bathroom?" and the husband grumbles, "I was just about to do it." This sense of having someone else say something just as you are trying to say it must is likely a frustrating feeling for anyone to experience.

In the case of engineers too, this gnawing feeling lingers for a long time. Engineers try to predict every possible situation, so they are always thinking about what to do next. All you have to do is trust them, and listen to them with a smile.

(3) Please do not allow the engineers get upset with you. But writing this, I find myself having a bit of trouble. I always make careless mistakes, forget promises and get my workplace all messy, and I get scolded by my boss all the time. Everybody was probably fed up with me.

The example I write here shows the type of careless mistakes that I would unconsciously make, but for the victims of the mistake, it could be a serious problem

It was at an exhibition, a few minutes before an important event to display and demonstrate our new product was about to happen.

As usual, the guests entered first, but one SONY staff member who had been leading the way came over to the display area, and he called to the engineer, "hi, thanks for all the hard work." But after that, he did something unthinkable! Without consulting with the engineer, he pressed the start button of the displayed device. The device had to be moved in a preset order, so it took at least three minutes to reset the system. We were really in a jam then. We had Morita-san delay his arrival, and all of the engineers had to redo the setup in front of our guests, which was a very disgraceful sight.

Has the record of no failures by the Development Team been broken by this event? Well, the exhibition itself went well, so I will let it go at this, but I was really dumbstruck by this incident.

WE MIGHT BE ABLE TO CREATE A "MAGNETIC CAMERA"

Right after we had released the Video Movie around October 1980, I thought that our next project should be none other than the magnetic camera that I had been dreaming of. I had a general idea based on the Mavica ,which I had been working on for a few years already, and the technique to record images onto sheet-type magnetic media was almost complete, and I assumed that I could make the CCD camera much smaller.

Incidentally, I have a beloved Muto drafting table in my office, so whenever I get an idea, I pull out some drafting paper and a pencil and draw up a plan right away.

I drew up the plan I had in my head, half enjoying the puzzle of floppy discs, batteries, motor, lens, CCD and other parts, trying to see if they fit nicely on paper.

I redrew it about three times, trying to make it smaller, and the last diagram was the camera that perfectly fit the image that I had in my mind. If I had been working alone like when I was younger, I would have drawn up the parts diagram all at once, but now that I had subordinates, I decided that this magnetic camera project should be run by Etsuro Saito. Just then, Saito-kun walked passed my office. I called out to him and invited him inside, showed him the diagram, and asked, "would you be able to make this?"

Around 1975, at the drafting table
in the development lab director's office

Saito-kun took on the challenge readily. At around this time, the Development Section 2 had many active managers in their thirties, including Sohei Masuda, Tadahiko Nakamura, Daiki Kouga, Minoru Morio, who was a year older, and Yasuhito Eguchi, Yoshiharu Matsumoto, who was a year younger, all of them being engineers who hadbeen pioneers of the video era. Saito-kun in particular was in a league by himself when it came to small parts and devices, so he was perfect for this project.

In about six months after that, he came up with a prototype he worked on with a few young subordinates.

A few days later, Vice President Ohga visited us, and seeing the magnetic video being used, he immediately informed Chairman Morita, who then ordered us to make five of the cameras and rush for the release.

When given such orders, we work day and night to accomplish what we can, but it can be difficult to gather some parts that need to be outsourced. CCD imagers and peripherals were not easily acquired. Saito-kun asked Ochi-kun to procure him several CCDs, but was rejected since there were none that worked without malfunction. He was at a loss.

As an emergency measure, I explained to Morita-san about our situation, and asked him to try to persuade Ochi-ku. That is how we were able to release Mavica. I am very grateful to Ochi-kun for the big favor.

Leaders are always going through enormous efforts in order to improve the image of SONY. Our newest developments are responses to these efforts, and once the objective and the schedule for something has been fixed, we follow them to a T. And if there is a better way to get something done, we will just let the leaders know, and they find that reassuring.

When we released the Mavica (Magnetic Video Camera) on August 24th, 1981, the surprise it gave to the media and the shock it gave to the world was immense.

In fact, the night before its release, a journalist from one industry paper was on a stakeout, waiting for me to come around to the front of my house, and he caught me at the front door. He wanted me to tell him what important product was going to be released tomorrow. "Aha!," I thought, and listened to his interrogation for more than half an hour, but since I continually answered evasively, he finally gave up and went away. Someone within SONY must have sensed the atmosphere and that something unusual was going on at headquarters, and leaked the release information to an outsider. Things like this happen frequently even now, so those involved should be aware (in other words, there are spies out there).

A truly surprising camera, the "MAVICA"

The release read, "Mavica is a still camera that mobilizes all of the new technology including the newly developed magnetic disc (Mavipack), CCD elements and IC technology. It converts images captured through the lens into electric signals by the CCD and records them onto the Mavipack. The images can be seen on TV right away using the viewer, and is also capable of rapid sequence shooting (60 shots per second) special effects. The signal recorded onto the Mavipack can be connected to a transferring VTR via a telephone line to use it as a color video camera, and also transfer it to the video tape to create a video album."

Nihon Keizai Shimbun wrote, "The camera and film industries find the product a threat to their labs since it requires no printing, though is taking a cautious approach regarding its practicality."

Asahi Shimbun wrote, "Images can be replayed immediately on home televisions, and negatives are reusable if the images are deleted. No changes or deteriorations in color. Impertinent to pollution."

Mainichi Shimbun had a different view. "Copies can be made freely without a darkroom, and the camera industry might get cranked up in visiting SONY for technical cooperation."

Weekly magazines usually do not write about regular new product releases, but with something that has as much buzz as the Mavica, all of the popular magazines were talking about it.

Asahi Weekly wrote, "The electric camera that SONY released eliminates the distinctive camera shutter sound. The soundless and filmless camera is called Mavica." This was the only media outlet that reported that the camera did not emit sound, and overall, comments were low-tone.

Yomiuri Weekly wrote, "Appearance of SONY's 'filmless camera' reveals bearishness amidst the bullishness of camera and film industries." The mixed responses of the camera industry were reported under this title.

Sunday Mainichi wrote, "A truly surprising camera has appeared… Media and Kabuto-cho (the business district) all amazed."

In addition, after SONY released its magnetic recording "Mavica," Asahi and Yomiuri, the morning papers of the day, posted the article on the top page, Mainichi on the top of the third page and Nihon Keizai Shimbun even attached a commentary and reported about it extensively." It is "exceptional" for an article about a new product to be posted on the top page, something that might happen just once a year. One publicist at an electronics manufacturer half-enviously said, "I was surprised to see the morning papers this morning. Common press rarely handles news about new products on their top page. I hear that SONY's Chairman Morita himself gave a big treat of running a demonstration himself at the press release, and it went very well. With that much coverage, the advertising would be worth a few hundred thousand yens. SONY is really skilled at advertising, but the media is somewhat lenient on SONY, so most articles about SONY tend to be treated as big news."

The fever cooled down in a few days, and bullish comments like how the Mavica is expensive or how the image quality is not so good, which would still not be much of a threat.

I agree, but you must not forget that technology progresses much faster than you think. If you take this lightly and lower your guard, you will end up regretting it. In fact, the pixels of today's CCD imager have easily exceeded the high-vision definition. The cost is unbelievably low, too.

Even in such times as those, most Japanese companies responded calmly and selected the paths that they should take in the future. Even with incomplete CCD imagers, they believed that they would be completed in the future, and never slackened on development. They challenged downsizing projects that seemed impossible, created technology that would make it possible, and reduced costs through mass-production enabled by mechanization.

WHY THE RUSH?

How was the response of Western countries? Though limited to only what I heard, I was often asked,

"What's the rush?" Everybody wished for co-existence and co-prosperity.

As I mentioned before, Western companies stick to working on things that capitalize on what they are good at and stay away from things other companies are working on. In other words, there is an atmosphere of respecting people's individuality.

As one such individual company, SONY brought unique products to the market and was gladly accepted by consumers worldwide. SONY had assumed they were co-existing and co-prospering. But SONY is a Japanese company, and in Japan, if you are not constantly on top, you could lose your spot to the competition immediately.

As such, the mindset of Japan and the West is fundamentally different, but neither is right or wrong. I think both are valid.

In Japan, companies' products constantly evolve based on the demanding principles of competition, so optimal products of good quality that are easy to use with minimal breakage are delivered to users at low prices.

These are products born from advancing technology that we engineers set as our goals. The sooner we complete a product, the faster we are able to please our users.

The answer to "What's the rush?" is that we want our users to be able to enjoy our new product as soon as possible.

After releasing the Mavica, we decided to launch a round-table conference with film, camera, and electronics companies as well as other related companies from all over the world to determine a standard format for the Mavipack (20inch magnetic floppy disc).

The format proposal was determined in 1984, and the Mavica was to be sold by all companies.

Later on, the Pro-Mavica MVC-A7AF was released in 1987 and was used mainly for reporting purposes by newspaper agencies. Since Mavica is an electric camera, once shooting is done, images could be sent immediately using telephone lines or a wireless connection to base stations or headquarters. In order to make this possible, Hidehiko Okada, under the direction of Tadahiko Nakamura of Team 3, developed a telephotographic device for Mavica. Asahi Shimbun offered its full cooperation in testing this device, so we were also able to create a color adjustment device so the colors could be printed clearly on paper.

In the Games of the XXIV Olympiad: Seoul 1988, pictures of competition results could be received and printed from Seoul right before the deadline of the evening papers, so Asahi was able to get of the upper hand on its competitors with scoop photos, and so the advantages of Mavica became famous all at once.

TAKE A PICTURE, THEN PRINT IT: "MAVIGRAPH"

Even when I was developing the Video Movie, I had digital print in mind. Video is for movies, and there are high-vision VTR, ENG VTR and Video Movies for 35mm, 16mm and 8mm respectively. So, our next project should be the digitization of still cameras and corresponding color printing technology.

I had to start by determining how best to print in color from a video signal. So in 1979, I asked Sohei Masuda, then Manager of the Development Section 2, to start running basic experiments on color printers. I also asked my subordinate Shigemichi Honda to experiment with printing technology, including ink-jets and ink-mists. I also asked Osamu Majima to experiment with the dye sublimation that he had come up with, which we decided was fairly good, and opted to proceed with full-scale development.

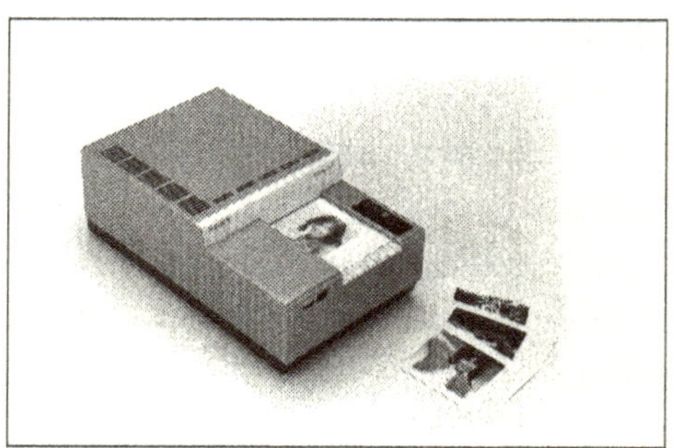

"Mavigraph", a video printer that created a stir
in the color printing industry

I chose Masuda-kun to be the project leader, Mitsuhiro Isogai in charge of the machinery, Daiki Kouga to develop the dye, and Yasuhito Eguchi to work on the electronics, thus establishing a highly capable development team. Conventional thermal heads that were available on the market then had huge fluctuations in the temperature increase of the elements, and were unable to obtain a uniform image, as images often came out with lots of lines. SONY's Sendai Plant excels at precision work like magnetic heads, so we asked them to produce a thermal head, and thus we were able to succeed in creating a practical thermal head after various improvements.

The Kanuma Plant, which is a mass-production plant for video tapes, had good coating techniques to create ink ribbons coated with different sublimation dyes, so we asked them to develop a practical ink ribbon.

It was our first time attempting to print a video signal in color, so the development consumed a lot of time and it was too bad that we couldn't make it on time for the release of the Mavica, but by the next year, in March 1982, we released the product as the "Mavigraph."

Being able to print video signals as a photograph with the Mavigraph created a stir in the industry by signifying an era in which color images could be easily created at home in little time with no pollution issues in the printing, developing or fixing processes.

When we released this product, I said something else that also created a stir. I said that digital printers like the Mavigraph will surely come to be used widely and popularly, and the time will soon come when people will be able to purchase it for less than 100,000 yen or even 50,000 yen. My subordinates who had been listening to me speaking in the conference room later said that they were startled. Because they knew that the cost of the product was 300,000 yen. Based on my experience of having developed dozens of new products, I was able to judge the future price of this product, or rather, its value.

If the product has high value, it will sell. If it sells, we can mass-produce it. With mass-production, the product can be made cheaper. Mavigraph was a product that had high value that could be sold in mass – so much so that it could turn a profit even if it was sold at 50,000 yen.

In recent industry papers, there was an article that reported about how demand for home-use color video printers was soaring.

"The biggest reason for the increasing demand of color video printers is because the digital video camera came about, the image quality of the printers improved and prices went down. In preparation for forming a full-scale market, SONY has released the CVP-M1 (78,000 yen), Victor Japan released the GV-PT1 (69,000 yen), Hitachi the VP-20 (148,000 yen) and another product, the CVP-W1 (25,000 yen) from SONY, with Casio and Fuji Film adding to the lineup."

As can be seen from this article, the 50,000 yen price that I projected might have still been soft. The times are changing fast. It is now a time that young people decide what products will come onto the market, and choose what products to buy. It is the era of electronics, and the era of digital.

The proportion of people who insist on old things is constantly decreasing, and younger people are creating new trends. As with the Walkman, CDs, MDs, pagers and digital tools, digital printers were also created by young people. That is the era that we live in right now.

2. HOIST YOUR SAIL WHEN THE WIND IS FAIR

NEWLY CONSTRUCTING A R&D CENTER

President Iwama said that he wanted to organize a new development department, and requested me to think of a name for the new team. My suggestion, the "Research and Development Center" was adopted, and the reorganization was executed in May 1982.

The former Development Section 1, Section 2 and the Display Development Sections were abolished and the Research and Development Center was newly established, with me designated as Director, Ohkoshi-kun as Deputy Director and Team 1 Manager. Horiuchi-kun was designated as Team 2 Manager.

This year, 1982, became an unforgettable year for me. Starting with the cosponsoring of the new format proposal for the 8mm video to related business circles in January, we released the Mavigraph printer in March simultaneously in Tokyo and New York, which marked the appearance of all of the final products of audio visual devices that I had dreamt of and worked so hard to create. In that sense, I deeply understand why President Iwama said to me, "Please work on a new direction of development from now on," when he established the R&D Center.

Right after the R&D Center was launched, in May, IEEE awarded us with the "IEEE David Sarnoff Award," which is often called the Nobel Prize of electronics. It was a great honor to receive this award.

But then, President Iwama, who had been undergoing treatment for a while, passed away in August after all remedies proved ineffective. It came as a great shock to all of us, leaving us all in a stupor, but we all took a vow that we would work even harder to carry on with the wishes of Iwama-san in the R&D Center.

NEW PRODUCT DEVELOPMENT STARTED

The former Development Section 2 had been working on crucial subjects such as the "Betamax,", "Video Movie," "Mavica," and "Mavigraph," all of which had been completely development and was passed onto Productions, and though there remained some backlog work, we could start the development of new products once again.

Since I have already talked about developments and topics relating to the Mavica and Mavigraph in previous chapters, here I will talk about something else.

Around this time, the word "New Media" was the new trend, and satellite broadcasting, cable television, tele-text and videotex (bidirectional communication using telephone lines) were being used.

It was Manager Nakamura and Kenji Nakano who made tremendous improvements in the performance of the video servo system by digitizing it and making it IC-compatible. And with the applied development of microcomputers, we converted the timer of the Betamax and the system control into microcomputers in 1977 and 1979, respectively. I asked Kosuke Komatsu, the staff member who enabled the high functionality of videos after Betamax, to work on the improvement of the videotex next.

The videotex was originally only capable of transmitting text and line drawings, and not natural images, but the automatic frame creator was generated by taking natural images into the computer and automatically extracting edges (determining the borders and regions of the image), coding it to dots and lines and polygonal commands and compressing 4Mbit data into one-250th. This new method was exhibited and demonstrated at the Chicago Videotex '84 in April 1984 and received favorable reviews. It went on sale in April of the following year as the videotex image input device Frame Creator System 1000 and came to be used by many consumers

The EBR, Electron Beam Recorder, basically creates a positive image by projecting electron beams onto the photosurface of mono-color 35mm movie film In order to create a color film, various techniques are required, such as taking three separate films of red, green and blue and putting them together into one color film, or creating a 24 grame movie film from a 30 frame television image.

The reason for the demand in EBR was because the High Definition Video System that SONY had been ahead of its competitors in developing went on the market in 1984. The image quality was praised as being comparable to movie films, and the movie industry made fervent requests to use the HDVS in movie production.

Why is it good to use the video in movie production? Because it can help reduce production costs. Movies need to be retaken quite frequently, and with movie film, it takes a few days to check the recorded film and redo the same sequence again later, so payroll costs for actors and other staff members could double or even triple. The HDVS can reduce these costs.

In addition, HDVS video can be erased and used again, unlike NG films, so expensive 35mm film is no longer necessary. Videos also have an established digital editing technology which shortens the time required for production, also reducing cost.

These are the reasons for the demand in EBR.

FOREWARNED IS FOREARMED

Naturally, the company requested that I be the one to develop EBR for the HDVS. I had always been thinking about learning about this technology, since I was sure that the time will surely come when recording technology for movies will become necessary. And my development specialty was in recording technology.

So, to be forearmed, I decided to work on the EBR. I found that the EBR developed by 3M was very well made technologically speaking, so I had it purchased right away in 1979.

Using this as a model, I asked Ozaki-san, a veteran staff member who can create anything movie-related on his own, to learn the basic technology of beam recorders, and had him develop a color film recording device for the HDVS.

After a few years of research, in 1982, we succeeded in recording with the HDVS. We made a preview room in building #200 and projected the image onto a 100-inch screen using a Victor cinematograph to evaluate the image quality.

The results were very positive, so we decided to produce a demo film, and with some help from Italian film company RAI, we produced a 10-minute movie film titled Arlecchino using HDVS video.

This film was released at the Swiss Montreux International Symposium in 1983, and greatly impressed those in the film industry. This system was acknowledged as a crucial tool in incorporating computer graphics images into movies, and is now used in SONY Pictures and SONY PCL in the States. In the future, I suspect that the demand for computer images will greatly increase.

SUCCEEDING IN THE DEVELOPMENT OF "VIDEO MASS-REPRODUCTION DEVICE (CONTACT PRINTER)"

After the Betamax was released, demand for soft tapes suddenly increased. Determining that SONY needed to develop its original contact printer, I resumed development in 1978, and in 1982, succeeded in developing a high quality printer using the magnetic transfer method.

The reason I succeeded was because a metal tape whose mother tape HC (High Coercivity) is three times higher than the slave tape became available, and because I created a mirror mother machine (mirror image pattern recording device) that creates a mirror image of the tracking pattern upon contact.

As expected, this method was of good birth, and in 1983, we were able to go into production by transferring this technology to SONY Magnescale.

In 1984, this product was released to the world under the name Sprinter. It was produced mainly for the Betamax, but in principle, it could be used for high-speed dubbing on VHS if the mirror mother tape could be created.

It could dub at 240x speed, and the mirror mother tape could be reused 7,000 to 10,000 times.

When in use, the mother tape is set in the endless box so it can be used for long-hour dubbing that can go as long as the pancake (shaped like a pancake) slave tape is.

This print system received favorable reviews, and it now holds a 30% share in the dubbing market, with more than 400 units in use.

THE "PERSONAL COMPUTER ERA" BEGAN

Exactly three years after the R&D Center was launched, our line of work had expanded. When we were still in Development 2, we had mostly worked on audio visual projects, mainly analog, but now, that had changed drastically and most of our development projects had shifted to digital technology.

Long before this, in 1964, SONY had developed and released the world's first all-transistor electronic desk calculator.

By that time, SONY already had digital engineers completing world-class ventures. Later, in 1967, the SOBAX ICC-5000, a device using full-scale IC was released, as was the ICC-600 the following year. Both of these showed healthy sales.

But the product concept of SOBAX was unable to keep up with the changing times, and production was terminated in 1973.

A lot of the engineers from that time are now exhibiting outstanding performance in digital product development, with Director Ochi, who contributed to the current prosperity of CCD, being one of them.

Most of the team, including Tadahiko Nakamura, Kosuke Komatsu, Tsunehiro Kashima, Yoshihiro Tanaka, Kenji Nakano, Hidetoshi Yoshimoto, Shunsuke Sakoda and Etsuro Saito from the mechanics division, all of whom worked on video developments and completed the digitization of syscoms, timers and servos, had transferred to my Development Section 2 and contributed to the development of SONY's digital products.

Thinking it would be a shame if the digital engineers cultivated by SOBAX were to be broken up due to the termination of the product, I recruited them to my team. I think this was a very good thing in retrospect.

In the same year that the R&D Center was launched, SONY released the micro-computer SMC-70, which has an interlocking feature with video devices. After that, full-scale personal computers such as the SMC-777/SMC-2000 were released. Before, people would crowd around computers such as the VAX-11 to type on the keyboard on the terminal, complaining about how slow it was, but now it was an era of one computer per person.

I started studying computers too, trying to keep up with the times, and I was creating programs using SMC-2000 at first, but then I found that the NEC PC-9801 had a quite user-friendly graph feature, so I was playing around making ray tracing images using C Programming Language.

DEVELOPING "WORKSTATION NEWS"

In May 1985, the President told me that he wanted to transfer Toshitada Doi to the R&D Center. I accepted, remembering it was he who had developed the CD. A few years earlier, Morio-kun and

Ochi-kun had received the Eduard Rhein Award, an award presented to people who contributed to the progress of television technology, for Video Movie development. Doi-kun had received that award too for developing CD signal processing technology and I was at the ceremony where he received it, so I knew that he was an excellent scholar.

Workstation NEWS

So I assigned him as the Assistant Manager of Team 1 under Deputy Director Horiuchi-kun, and also as the Manager of Team 3.

Incidentally, the Manager of Team 1 back then was Hidehiko Okada, Team 2 was managed by Kosuke Komatsu, Team 4 by Shinji Amari, Team 5 by Kunihiko Arai and team 6 by Kanji Murano. The R&D Center comprised of a total of 155 staffs.

Hori-kun, Tanaka-kun, Hayashi-kun, Tezuka-kun from the MIPS Business Division managed by Nobuyuki Idei (now President) and seven R&D members, a total of 11 staff members were assigned to work on the 32bit engineering workstation computer.

The project was named ICKI by Doi-kun, meaning chugging a drink in one go.

We produced a prototype for this project in February of the following year in just six short months. By May, the Super Micro Business Development Team had been established, and ICKI was handed over to Productions.

HANDLING "THE SONY RANKING"

The name of the development team I managed changed from time to time, but at the end of the "core tasks," there was always the phrase, "other special assignments."

This was so that we would always be ready to work on urgent projects, experiments or developments given to us by Ibuka-san or Morita-san. The leaders are all impatient (in my opinion), so my team, which had engineers who were able to take action immediately, must have been convenient for them.

This time, Masaaki Morita (then President of SONY Corporation of America) suddenly asked us to "create a personal golf ranking system using computer software that the PGA (Professional Golf Association) is proposing and have the whole world use this SONY ranking starting from the fiscal year of 1989."

I immediately had our software engineer Hiroyuki Kobayashi create the program, and installed the entire system at the IMG (International Management Group) located in Saint Andrews, England and started operation.

If you play golf as a hobby, you have likely seen golf magazines that say "according to the SONY Ranking, Ozaki ranked 9th at an average of 10. 82 points."

This project was handed over to the Kihara Research Center, but the system was upgraded every two years or so, and last year, Takahiro Ishii made major changes so that it could operate on Microsoft Windows. "The SONY Ranking" renewed all of its features and is transmitting information all over the world with SONY's name attached to it.

THE "KIHARA SCHOOL" WITH BIG DREAMS

I am not sure when people started calling it the Kihara School, but it was probably when I received the Science and Technology Agency Director-General Award and all of my former subordinates got together for a toast. Someone said, "Having learned my job under Kihara-san, I am able to enjoy my working days now. It was like a school to teach us all of the techniques." Since then, this "Kihara School" became something like a symbol of solidarity.

Inaga-san, who transferred to Productions with the PV VTR, Kato-san with the CV VTR, Horiuchi-san with the U-matic, Fujiwara-san with the broadcasting VTR group; all of the people who cultivated their skills under me are alums of the Kihara School.

Since then, the outstanding alums of Kihara School is supporting the core of SONY from Beta to 8mm, printers and computers.

I believe that there was not one single dropout, and I am very grateful for that. But some have come back, like Horiuchi-kun, who boasts that "I'm the boomerang of the Kihara school." These veterans who came back after experiencing production and marketing are valuable resources in the R&D Center.

In the Founding Prospectus that Ibuka-san wrote when he established the company, one of the objectives stated is "To establish of an ideal factory that stresses a spirit of freedom and open-mindedness, and where engineers with sincere motivation can exercise their technological skills to the highest level." The Kihara School has practiced this firsthand.

Development section 2, which had been dubbed the Kihara School, always maintained an environment in which engineers could work on development at their will, under my policy of "developing technology while training personnel."

Thinking back now, I truly believe that I was able to train serious and pure-hearted engineers, having them learn technology mainly on recording, in an environment that was free-spirited where everyone could enjoy work in unity and harmony

The business policy states to "We shall eliminate any unfair profit-seeking practices, constantly emphasize activities of real substance and seek expansion not only for the sake of size." So I focused on the substantiality of technology. I never worked on anything that lacked content and devoted myself to only practical work. SONY became a major company only as a result of these commitments and never unnecessarily pursued scale.

"We shall be as selective as possible in our products and will even welcome technological challenges. We shall focus on highly sophisticated technical products that have great usefulness in society, regardless of the quantity involved. Moreover, we shall avoid any formal demarcation between electronics and mechanics, and shall create our own unique products uniting the two fields, with a determination that other companies cannot overtake." This was the development policy assigned to me.

Yes, difficult development is much more rewarding. High-end products make a much better show. Coming up with ideas integrating electronics and machinery is the perfect job for a mechatronics specialist like me.

Thus, it is only natural that SONY would be able to get so far ahead of its competitors and create the first of a given product in

Japan, or even in the world.

I hope that all of the engineers who have inherited and are practicing this policy of Ibuka-san's will advance onward, overcoming any barriers they may encounter.

I also hope that you understand the real meaning of the Founding Prospectus. I expect those who understand Ibuka-san's spirit and have inherited it to build SONY in the new era and lead it to success.

MY "LUCKY THEORY"

By the way, I have always told young engineers,

"Don't fret over your failures. Think of it as being lucky."

All engineers progressing in their careers will run into a wall at one point or another. The more serious the engineer is, the more he will become exhausted and depressed. Whenever that happens, I give him a pat on the shoulder and tell him,

"You have nothing to be depressed about."

He will look puzzled for a moment. Some even repel, saying "how can you say I am lucky when I have made such a stupid mistake?" But you should think harder. Failure is the best experience. You experienced firsthand what would go wrong via your own actions. So next time, you can avoid this path. I call this "the lucky theory."

I have lived a very blessed life. What underlies that life is my confidence that I have learned from all of my failures.

There is a saying that goes "out of disaster springs fortune." That's right, you don't have to be bothered by your failure forever. It will only make you sick. You should continue doing your job constructively with a positive mindset.

A long time ago, Ibuka-san told me that human beings have the capability to heal themselves. However, excessive stress causes various type of hormones capable of causing ulcers and cancers to be released. This can also be controlled by one's mindset.

According to today's medicine, the brain releases two different types of hormones: beta-endorphins and adrenaline.

Beta-endorphins work like morphine, helping to maintain youth, improve immunity, kill cancer cells and keep people in a happy mood.

On the other hand, adrenaline is also a toxin several times more poisonous than snake venom and can contract blood vessels, increase blood pressure and cause strokes, dementia and cancer.

Based on my Lucky Theory, let's think of tips to make you healthy incorporating this knowledge.

Whether it be company performance or research and development results, our ideas are almost all linear, so our daily tasks proceed in a predetermined pattern.

However, reality happens in a nonlinear manner, and unexpected events happen one after another, throwing you off. This causes stress, releases adrenaline and makes you sick.

If the world is unpredictable, you should not get upset about whatever happens, and instead just look for a good way to solve the problems whenever they arise. This is the Lucky Theory. And this is my tip for leading a healthy life.

I generally prefer not to write personal facts about myself, since I am modest and shy, but as a way of atonement for those who have read this book that demands you do this and do that, I will write a just bit about myself.

- Whether it is work-related or not, I get deeply involved in whatever I find interesting.

Electronics, machines, physics, chemistry-- whatever it is, I get visceral in making things, adjusting them and experimenting with them. In a sense, I might have been a very amateurish person.

- Not only do I get interested, I try to do things on my own, experiment on my own, and create products on my own.

I never stop until I am satisfied. I never give up, even if I fail at times. I repeat things over and over until I am able to successfully make a product. This must be due to my craftsman spirit.

- My point of view.

Whenever I am walking around town or riding trains and airplanes, I rarely do so absent-mindedly. I always find something interesting and start thinking about it.

There are many examples, like: when you are on an airplane flying at night and look out of the window, why can't you see the Milky Way? I have seen the aurora three times near the North Pole. Did you know that when the Concord is flying at mach 2, the metal frame of the door gets so hot that you could get burned? I also won a quiz, which required you to notice the pamphlet on your seat and listen to the announcements. I won a champagne and a bag. Twice. These are examples regarding airplanes.

When I had overseas business trips, I would always notice new products being developed. It must have been my keen eyes that found clues here and there. One of them was the finding of the chromatron tube.

Regarding know-how, the "hoist your sail when the wind is fair" I talked about earlier is the accumulation of know-how.

If you can do something and succeed, you have the know-how. That makes you better than other people. It means that you are the only one who can go ahead with it, and you will accumulate know-how. Then, you will eventually build up a pyramid of excellent accomplishments that no one else can compete with.

In other words, you should take good care of your know-how for things you are good at, nurture it well and advance with it.

I will also talk about how "colds can be useful depending on how you suffer them." I always overwork myself, but while I am concentrating on a project, I never come up with ideas for something else.

Whenever I come up with new ideas, I need a different space.

In my case, that space was in my bed when I had a cold. In essence, it's fine to be busy working on something, but you need to take a break once in a while and think about something you hadn't had the occasion to think about. When I was forced to take time off due to catching a cold, I would quite often come up with good ideas while resting.

I say again here, the important thing in coming up with good ideas is to never be a copycat. When SONY was still Totsuko, we learned from others in developing our products for the first decade. But after that, our slogan was "never copy people. Let's make something better." We can make new products only when we stop copying others. But it is difficult not to copy people in the world of technology. People might catch up with you while you are working on something new so there is always pressure to hurry up and finish. It is easy to copy something that already exists, and t may help you to take over the company that made that original product. But the engineers won't learn the know-how regarding that product, and won't be able to learn the skills needed to create the next new product.

"KIHARA RESEARCH CENTER" ESTABLISHED

Following the words of Morita-san (then Chairman), to "continue free research at the new Kihara Research Center," I thought that the best way to repay all of his kindness was to develop the kind of new technology that SONY would need in the future. So I decided to develop a world-class system of three-dimensional computer graphics that SONY headquarters had not explored at that time.

This is how the Kihara Research Center was established in October 1988.

● The springhead of the Kihara Research Center

[Charter] Work on the research and development of creative and unique products in a free-spirited environment in order to contribute to SONY and society.

[Main Research Projects]

(1) 3D computer graphics

(2) Video image real-time processing

(3) Parallel computer development

(4) Computer software program development

At the time of establishment, the company was managed by six board members.

Representative Director: Nobutoshi Kihara

Managing Director: Kosuke Komatsu

Director: Minoru Morio

Director: Teruaki Aoki

Director: Masayuki Takano

Auditor: Akihisa Ohnishi

The Kihara Research Center started with 12 members assigned to it by SONY, but I decided that, in order to study computer graphics, we needed to invite an experienced external engineer. Since then, we worked very hard to expand the number of proper employees every year. Due to this effort, we succeeded in gathering high-leveled engineers in both the fields of hardware and the software. As of 1996, a total of 38 elite engineers are efficiently proceeding with their projects.

The company is comprised of two departments, Development and Planning. The Development Department is led by Managing Director Komatsu with Manager Hiroshi Hayashi for Team 1, Manager Shinsuke Koyama for Team 2, Manager Naosuke Asari for Team 3, and Komatsu-kun leading Team 4. The Planning Department is managed by Manager Tsunehiro Kashima and Officer Kaneyuki Kuragata, our two veteran engineers. Michiko Tsukamoto assists me as my secretary.

I believe that our mission statement of "contributing to SONY and the society" is being steadily achieved.

LOOK FOR SOMETHING THAT IS "OF GOOD BIRTH"

The Policy of New Product Development

THINKING ABOUT NEW MARKET DEMANDS

There is nothing particularly new about the policy of new product development. You may think it is just a list of normal things.

And that is exactly right. All of these are perfectly normal to engineers, but if any one of them is forgotten in the line of work, various things – ineffective things, unnecessary things, things that are not new, things that are simply copies, things that do not relate to future progress – can become factors of failure. This I can say from my own experience.

When the tape recorder was developed, it was only purchased by courts and for other specific purposes at first, so Morita-san wrote a book titled "Tape-style Magnetic Recording Device," which became a best-selling book that got reprinted up to its 32nd edition and also ended up developing interest among the public. As a result, people in various fields understood that tape recorders could be used in various ways, and the G-type enjoyed healthy sales, while at the same time, users started to make new demands on tape recorders.

After the tape recorder, it was the transistor radio. If the radio was transistorized, it could become portable, able to fit in your pocket. We figured that would be very convenient, so we decided to transistorize it.

The same goes with the TV. If we could transistorize it, it would consume less power, and we would be able to make it portable. Small is good, we thought, so SONY has always thought about downsizing its products. I think that this idea of downsizing took root in SONY during the era of the transistor radio. It was all because there was a demand from the market, and the company responded to that demand by making efforts to downsize its products.

We made an open-reel VTR CV. But the market was not satisfied. Instead, they kept urging us to make it cassette-based. SONY's engineers worked day and night to make a cassette, but soon the market said, since the TV is in color now, the VTR should be in color too. It took a long time to add color to the VTR. Then, we needed a portable color unit.

The engineers had expected such demand and had already been researching it. But the pressure to do this and that was much bigger than we had expected. We engineers had to work on development as if our lives depended on it. If we were to say, "no, that's impossible" to the demands of the market, development would never progress.

We rather welcomed the pressure, and when the pressure was on, we refused to give up. We were full of passion, everyone working in unity, so development progressed at an amazing speed.

Whatever the market demands, remember that "the customer is king." Whatever the king demands becomes a necessity. You should work on developing something that will make the king happy.

PRODUCTS EVOLVE THROUGH FEEDBACK LOOPS

Ibuka-san and Morita-san found a new market opportunity in tape recorders, and the information on relevant issues and market demands were passed on to me right away, so I was busy handling those demands.

Fine details such as power voltage fluctuations, tape-speed changes and attribute issues could be handled on a case-by-case basis by thinking of the best measures to take, but fundamental issues such as downsizing, lightweight solutions, cost reduction and making it portable required a totally different concept to fulfill demands. We just kept on developing and developing in order to meet the challenge of these tough demands, and we felt a sense of achievement and satisfaction when we could solve the problems.

Products evolve. Marketing and Development has maintained good relationships to form a solid feedback loop to allow things to evolve into the next new form, repeating this over and over again until another feedback loop is generated. This process is what led to video recorders and magnetic sheet recorders, each f these further growing into new categories.

If the feedback from Marketing stops, or if Development fails, the loop is broken, and products will cease to evolve and die off. We must be careful not to cut this loop.

A growing loop will quite often generate other routes. The VTR was derived from the tape recorder, but the details and the technology is basically the same and the know-how has been passed on as a whole so engineers would be able to continue to work in the new field as VTR engineers. This is a very efficient route.

For example, when we were thinking about the consumer VTR CV-type, we employed the field skip technology, using only one motor and a rubber belt brake servo, eliminating power transistors, mechanically generating the synchronizing signal to operate the vi-di-con camera. In order to reduce cost, we came up with other new ideas and put them together comprehensively, which resulted in a device that surprised the entire world.

I often observe engineers who, when thinking of a new method, get excited over just one idea and focus solely on that, but in most cases the results are not so great and the product just fades away.

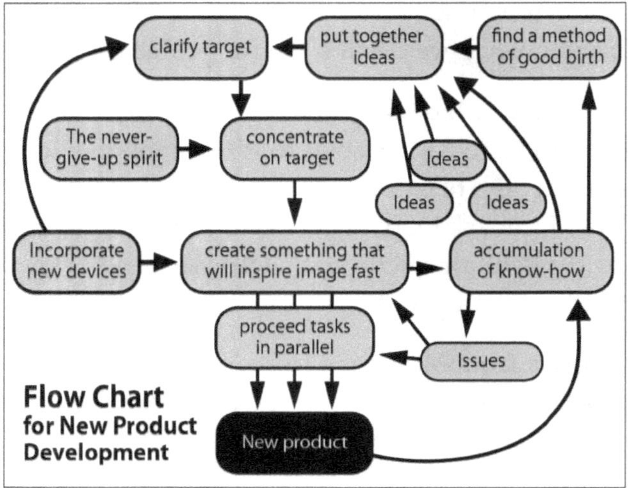

There is the case of RCA, the fixed-head VTR that Japanese companies released all disappeared once the CV-type was released.

The 3/4-in. cassette four-head VTR released by RCA persisted only because it was easy to insert cassettes, but it did not solve the issue of downsizing, so this product faded away too after being put into competition from the U-matic.

Even the VCR VR-2000 by Philips was terminated for having stuck with the double-sided cassette and eight-hour replay, even though its production costs increased and malfunctions happened often.

The reason for the cost increase was because they made the track pitch thinner to replay the tape for eight long hours, then employed a bimorph variable servo for tracking, and the development proceeded in the wrong direction. I think this happened because they misjudged whether the method was "of good birth" or not.

FINDING SOMETHING THAT IS "OF GOOD BIRTH"

Whether it be machines or electric circuits, or any other parts and materials, everything must be "of good birth." As might be expected, if even one small part fails, the entire device will become a useless defect. Parts can be replaced and repaired, but systems and methods cannot be replaced after the product is completed, so it is crucial to carefully consider everything conceivable from the beginning of production. This is what I mean by finding something that is "of good birth."

It might sound natural, but what is "good birth?" And how should you seek it out?

When we were putting the VTR in a cassette, for example, we had many ideas. Some may be great ideas, but you shouldn't be excited about this. You should think, "if I think harder, I might get an even better idea."

Of course, we designed, built and experimented with all those ideas, which resulted in more than ten prototypes. Whenever possible, we developed them in parallel to save time. This led to some of our prototypes being improved and upgraded twice or even three times. A year later, we finally completed the "U-loading" system.

Among the 10 different loading systems, we judged that the device that functioned most smoothly during assembly and adjustment was the best loading system of good birth, and chose that one. Incidentally, this concept of a cassette loading system is now used in all VTR including Betamax and 8 mm, which proves that it was definitely a system of good birth.

Devices that functioned smoothly are ones that had no big issues from the designing stage, did not require high precision in its parts, did not require expertise in assembly and adjustment, and once you built it up and turned it on, it operated immediately. This is what you call a device of good birth. When a device like this is taken to production, the reasonable design enabled a sensible mass-production. So, being of good birth means that it can produce reliable products.

TO BE ABLE TO JUDGE WHAT IS "OF GOOD BIRTH"

It is not so easy to be able to judge what is "of good birth." You cannot learn judgment just by listening to people, so you must acquire know-how from your own experiences and accumulate know-how over a long period of time, after which you will finally be able to judge whether something is of good birth or not.

When designing a device, try to draw a plan of the design you think is most appropriate for the purpose, and build the parts yourself. If you can build it up smoothly, check the movement. Did it work on the first try? Great. That doesn't happen often.

Then, modify the design to make improvements. If the modification doesn't work and you have to modify the design over and over, you won't be able to produce a device of good birth.

For example, there are VTRs on the market that have a tape path that was badly designed, causing unstable strength to be applied to the back tension. These devices are sold with an impedance roller to prevent the tape from squeaking. This is one example of a device of bad birth, and having to sell a product with something that should not be necessary on a usual VTR (a plaster to cover the injury) is a pretty bad situation.

Electric circuit designs can't be that bad, but there were cases that "plasters" to prevent rush-in noise and transmission were applied. However, electric circuits are now mostly IC or LSI, so they are much more stable.

We have no knowledge when we are born. We absorb knowledge from our parents, teachers and seniors while growing up and become an adult. All knowledge learned in the meanwhile is mostly memorized, so we train to recall what we memorized. This is another very important type of training. Abundant knowledge comes from these memories. Just as we refer to things like "the well of knowledge," the bigger this well gets, the more chance there is for the knowledge therein to come in handy. In other words, knowledge exists in parts. Parts that, when gathered and assembled, become a wonderful product.

ENGINEERS NEED SOMETHING MORE THAN JUST KNOWLEDGE

Whether you are a liberal arts person or a science person, the knowledge you learned in school will come in handy in real life. But for science people, I think you need something more than just the abundant knowledge you have acquired.

By "something more," I mean problem-solving skills. The algebra that I learned in school was exactly that – how to solve problems. This was not a class to simply memorize things, but rather, to think of the meaning of each step and take those steps one by one to get closer to the solution. Of course, it is often easier to solve the problem if you have memorized few equations. But equations are parts of knowledge that your predecessors created for you, and these parts need to be assembled as they alone cannot nurture your ability to solve problems.

Perhaps the reason why people who like math choose to study engineering is because they feel that a workplace which allows them to work using their problem-solving skills best suits them.

I am one such person. I loved algebra when I was a student. The reason I loved it was because you could always find the answer. In algebra, you can calculate backwards to check if you are right, so I almost always got full scores on my tests. I was debugging my careless mistakes. I felt so good when I could solve a problem, and felt like challenging the next difficult problem. This was a feeling similar to what I feel when the device I developed worked as I had expected, and the feeling to want to make it even better.

LEARN BY GETTING YOUR HANDS DIRTY

I would like to explain how to learn problem-solving skills, which are crucial for engineers.

Even when someone teaches you know-how, you will be unable to grasp the essence of it. Real know-how can be learned only when you have experienced and understood it.

No matter how much you read great books, the author's know-how will rub off on you. It will just stay in your head as separate parts if you just read it and say, "oh, okay, I got it." So, how can you learn the know-how?

Try replicating the author's achievements. It might not be possible depending on the content, but try replicating even just a part of it and you will surely learn something.

This will become a part of your know-how. What you absorbed as just a part when reading the book will be accumulated now as know-how.

When I design a machine and draw up the plan, I turn the lathe and build the machine myself. I build everything on my own, and if I find something that doesn't work, I remake the parts myself and fix it until I am satisfied. This way, my own design errors and improvements are accumulated one after another as know-how so I will never forget them. What you learned by getting your hands dirty becomes your own valuable property. You will never forget that know-how.

SONY has always produced devices that created a buzz, so we never copied others. Ibuka-san hated copying. But some manufacturers do just that. They even make the external design of their products exactly the same as ours, which leaves me dumbfounded, but as long as they are only copying others, their engineers will never be able to accumulate know-how, so they can never produce groundbreaking devices. I feel sorry for those engineers.

CREATING REASONABLE DESIGNS

I always say to my subordinates, "Machines are devils." Just making things work isn't sufficient. Machines are causing problems in microscopic parts where you can't see.

For example, there are lots of problems that occur in the ball bearing. Somehow, it generates power! Then the rotating head bearing picks up the generated electric spark, and noise occurs on the monitor. In order to remove this, I had to attach a brush terminal as a ground terminal on the rotating shaft.

There are also cases when the images replayed on the VTR sway. This too happened because of the ball bearings. When the diameter of the balls was irregular, the rotating drum axis shakes and the image sways. It is not sufficient for machines to be just operating. Even a single micron of error affects the images.

When using ball bearings in VTR, various problems occur. But if you blame it on the ball bearing and forcibly fix the ball, noise could reappear over time. Images could sway. Fixing the bearing may not be a radical solution. Rather, you should think about eliminating the cause completely, and choose to attach a ground terminal. You should not blame the bearing for the image sway, you should use two bearings to apply lateral pressure and eliminate the shaky bearing. I hope that you now understand what I mean by reasonable design.

ELECTRICITY IS HONEST

Electricity, on the other hand, is honest. It functions smoothly and as calculated according to the circuit diagram. If an experienced engineer with know-how designs it, it will almost always function without fail.

With the IC and LSI digital circuits today, debugging and simulations in the early stages are a lot of trouble, but once you complete the circuit, it rarely malfunctions and lasts a long time, which is a great advantage.

As such, electric technology is ever-evolving. If you let your guard down, past technology gets old very quickly, and no one will buy it.

One example is the old tape recorder and VTRs. Looking at the cost structure, machines took up 80% of the cost, while electrics took up only 20%. But now, the ratio is reversed, and it is 20% machines and 80% electrics. Why is that? Machines used to cost less than electrics in the past. So engineers set machines as the core in order to make things cheaper.

But now, all faults in machines can be fixed by electrics. For example, in the past, the mechanical section of tape recorders and VTRs used rubber to transmit power and convert rotations, but rubber creates unstable slips and was short-lived, so it was eliminated from the power transmission and was replaced by brushless motors or transistor servo motors.

By replacing mechanics with electrics, a more stable and long-lasting device could be built.

TV tuners became digital tuners, timers became crystal, tape counters became liquid crystal, wired controllers became infrared remote controllers and zoom lens used linear motors.

Choose methods that match the times. Rather than forcibly solving problems with machines, use electric parts if they can be used as replacements.

COMBINING SECURE SYSTEMS

In my experience, I hardly ever succeeded when I used parts that I had never used before, or forcibly used some mechanical system on a whim that had never been mass-produced.

Things without accumulated technology are highly likely to have some serious drawbacks somewhere. Do not use such things, but rather, use mechanical technology that has been tested over time, or machinery that had all issues pointed out by users addressed, and you should be able to use machines without problems.

When I was trying to design the home-use VTR, I used the open-reel tape recorder TC-101, a device which was said to be an excellent machine as the base, and attached a video rotating head on it. By using a proven tape recorder machine, the CV VTR never had problems with running the tape. That is how we were able to release the CV VTR to the world as the world's first home-use VTR, and rapidly grabbed the spotlight as a mass-produced device.

With the U-matic VTR, I incorporated a new tape loading mechanism onto a device that had good records with the CV VTR, and it showed success as the world's first cassette VTR.

Then, based on the loading method of the U-matic VTR, we made a full-scale home-use compact cassette, the Betamax. The Betamax is the antetype of home-use compact VTR, and became a breakthrough product which created a VTR boom all over the world.

THE NEVER-GIVE-UP SPIRIT

Then came the 8mm VTR. After the Betamax was released, the ultimate video which used the best technology was demanded, and as I explained earlier, I thought about developing something that was half the size of the Betamax, and established two teams to compete in the development process.

It is necessary in development too to apply the competition principle in order to achieve your goal as fast as you can. I was observing the progress, and one team changed its course, assuming it was impossible to half the product. They altered the reduction ratio to 70% from 50%, and submitted their product. I was very disappointed by this. Their mission was to make it 50%. A 70%-reduced in size product did not seem small at all. That is natural. Human eyes sense that half is very small, but with 70%, the drum diameter did not appear to be much smaller compared to the VHS VTR. This was not good enough to be produced as the next product, so I disbanded that team.

So then, how did the other team do? They understood the meaning of the "half the size" project, challenged the barriers, overcame them one by one and completed their final target, the portable video movie.

Looking back on the history of halving the recording wavelength, starting from 1965 when CV VTR was released up to 1975 when the Betamax was released, ten years have passed. And the first 8mm Video, the CCD-V8 was released in 1985, so incidentally, another ten years had passed before the "half the size" project succeeded.

After we released the video movie, we hosted a standardized format conference with companies all over the world and worked hard for several years to establish a world-standard 8 mm video. The 8 mm development surely succeeded because we spent years of hard work, because we had the never-give-up spirit. The passport-size product enjoyed explosive popularity, and established the digital video era.

TAKING ADVANTAGE OF LEADING-EDGE TECHNOLOGY

Needless to say, if you let your guard down, people will get ahead of you. The point here is to know a good thing when you see it. Also, remember what happened in the past. Even if there are methods, materials or ideas that you were unable to make use of in the past, they may come in handy sometime in the future. Servo systems, azimuth recording in the Betamax, head material and magnetic tapes are some examples of those. Azimuth recording was an idea that existed back in the time of tape recorders, in fact, and I had kept it as an aside, thinking it might become useful one day.

The same goes with the head. Hard materials that could not be used when we had no processing technology, or materials that were difficult to process could now be used with the progress of processing technology. Magnetic tapes could be used for the same reason. I once developed an HID tape for the tape recorder. It used alloy metal powder instead of ferric oxide. The idea was fine, but since it was agglomerate powder, the acicularity and squareness was not ideal, and it could not be used for video. But now, it is used in a different form.

Like this, you can look back on past technology, and there are many things that can perform greatly even now. There is no way you should overlook this. I always have my antenna up to be the first to find new processing technology, new material, new methods and to take advantage of them and create new products.

Products that were completed always came with titles such as "world's first" or "high performance." With the development policy I have, I believe that I can definitely create great products.

TIMELINE OF SONY'S TECHNOLOGICAL DEVELOPMENT

> **Era of absorbing and accumulating knowledge**
>
> - vacuum tube era

1950s

Tape recorder G-type released

Portable tape recorder

Stereo tape recorder

Semiconductor production starts

Transistor radio released

Transistor VTR research starts

> **Era of Building up Original Technology**
>
> - semiconductor era

1960s

Monochrome transistor TV research starts

Compact VTR PV released

Philips method compact cassette recorder released

Hi-fi acoustic equipment
 -- amplifier, microphone, speaker released

Trinitron color TV released

Era of High Performance, Downsizing and Mass Production

- ultra-precision work era

1970s

Industrial VTR U-matic released

Broadcasting VTR Omega-type released

Home-use VTR Betamax released

3. 5 inch floppy disk released

Walkman released

Era of Multiproduct, Automated Mass Production

- optical application era

1980s

CD player released

All-in-one camera 8mm VTR released

Digital camera Mavica released

Digital printer Mavigraph released

High-speed assembly robot released

Workstation computer NEWS released

Communications related devices and telephones released

HDTV (high definition TV) released

Looking back to the timeline of SONY's technological development, you will notice that great changes came every decade.

● Era of Absorbing and Accumulating Knowledge

Prior to this era, the war had left a huge hole. We had practically lost everything.

The entire country was at a loss, and everyone was searching for what to do. The only way to acquire technology was to gather information from developed countries in the West and study them, or have them teach us.

That is how Totsuko started studying tape recorders, by gathering information. Engineers enjoyed their work so they worked night and day, but the factory was strapped for cash, and though we were receiving salaries, they often came late.

Overcoming such difficult times, the tape recorder started to sell, and we proceeded to the next leap forward, and carved out the new era of semiconductors.

But there was still much to be learned from developed countries in the West, and this decade had to be spent absorbing their technology. Regarding peripheral technology in particular, the quality of our parts and materials was far inferior to those of the West. We had to import resisters and condensers for electric parts, and I recall that it was a long time before Japan became capable of producing compact solid resisters.

Even with a 1mm-thick steel plate used for base chassis of machines, Japanese plates were too blunt and weak, so we had to use a slightly thicker plates, which made the machines heavier in addition to many other disadvantages.

Even with the setscrew, Japanese screws were so weak that if you tightened them too much, they would be wrenched off. It was much later that the Philips screw (cross-slot screw) came to be widely used.

• Era of Building up Original Technology

In this decade, the 1960s, we were able to make use of the technological know-how accumulated in the past decade with our applied transistor technology.

Not only tape recorders, but TV and VTR were also transistorized. Transistors for high-frequency, which were initially thought to be impossible, were enabled by the untiring effort of semiconductor technology in VHF and UHF tuners, and even SHF 12G Hz was made possible with semiconductors.

SONY has created any type of semiconductor, however unique it is, as long as it was needed for our own products. Even back when we were producing tape recorders, we would produce our own optimal tapes and heads, and that is how we are able to create a device that can perform at its best. That had been our belief.

Ibuka-san would stress this to visitors, especially those from overseas. "SONY is the only company in the world that produces tape recorders, tapes, heads and even motors in-house. The same goes for semiconductors. That is why we can produce high-quality products."

Our original color TV and the ultimate color Trinitron appeared at around this time. It was natural that SONY's strenuous efforts bore fruit in this era.

• Era of High Performance, Downsizing and Mass Production

In the 1970s, as it is replaced the era of ultra-precision, most of SONY's products were mechatronics products such as video, Walkman or floppy disks.

This was a time when videos were in pursuit of high-density recording, so naturally, compact devices required precision processing and dimension accuracy. Machine tools capable of precision processing were thus needed, and therefore precise electric measuring tools were developed, and the processing accuracy of many machine tools improved by more than ten times.

Sony Magnescale, the company that was established in 1969, was a company that produced and sold measuring tools, scales that used magnets.

Next, as a scale of video precision, I will show you a table comparing recording density. Video development was a battle with high-density recording. In particular, we had no choice but to improve the machine precision in order to reduce track width, so naturally, machine tools evolved to become ultra-precise.

It was in 1959 that the U-matic global standard was defined, and since then, the amount of video tapes was cut down by ultra-precision processing.

Purpose	Name	Tape width	Speed per second	Tape volume
Broadcasting	Four-head	2-in.	380 mm	1/1 for comparison
Broadcasting	Omega	1-in.	190 mm	1/4
Consumer, educational	CV, AV	1/2-in.	190 mm	1/8
Broadcasting, industrial	U-matic	1/4-in.	95 mm	1/10
Consumer	Betamax III	1/2-in.	13 mm	1/117
Consumer	8mm video	SP 8 mm	14 mm	1/172
Consumer	8mm video	LP 8 mm	7 mm	1/344
Consumer	Digital video	DCR-VX1000	6. 5 mm	1/344

This was the time when both video and TV went into mass-production. In addition, tape recorders, represented by the Walkman, became an explosive hit, and methods to produce high-performance devices by assembling tiny parts were established during this era.

SONY became a byword of compact products, and it was probably from this era that Japanese home electronics started to be called "light, thin, short, small."

● Era of Multiproduct, Automated Mass Production

The 1980s up to today can be defined as the era in which we handled various products while at the same time saving energy in mass-production, by automating almost all processes from making parts to assembling them. This is not limited to the home electronics industry, and automated mass-production is becoming the global norm, including in the automobile industry.

One unmissable part of technology in the home electronics industry is the optical application. Semiconductor lasers have become commercially viable from this era, but considering the fact that CDs have virtually driven out vinyl LPs in less than five years, optical application products should be invented one after another, soon penetrating the video industry as well.

Further, expectations are high in the commercial viability of blue semiconductor laser, and I predict that the future of optical technology will develop even further.

● So, what next?

Based on the 50 years of development that I experienced, everybody has asked me "what's next? " but my answer is always, "I don't know."

Most of those who ask me are just being playful, thinking "Kihara-san is always making strange things, so I wonder what he is thinking now." They probably think it might be a good topic of conversation if they could get something out of me.

Surely, I was thinking very hard about the next thing. So, I couldn't say anything. But when I got an idea of something that made me think, "This is it. I will test this next," it gave me the shivers. Then, I would draw up the plan all at once, connect the electric circuit, and create something new. That has been repeated up to this day.

New technology and inventions are born by combining existing ideas and materials, so they evolve over time, repeating a cycle of progress and regress. During this time period, there will be various external factors. If there are wars or economic fluctuations, the world may take an unexpected turn.

In such a nonlinear, super-real world, it is basically impossible to predict future events, much less the future of technology. But though unpredictable, it is important to have dreams. Ibuka-san told us about many of his dreams, and everyone got together to realize those dreams, one by one, which built up the foundation of today's SONY.

So my conclusion to the question "so, what next? " would be this. I can give you an answer for a future that is a year or two away, but regarding SONY in 30 years or 50 years, I believe that great outcomes will be achieved if all SONY employees hold many big dreams, brush up on and accumulate their technology through untiring effort until their dreams come true one by one.

CHRONOLOGY OF NOBUTOSHI KIHARA

Year	Biography
1926	Born in Tokyo, October 14th
1947	Graduated from Waseda University Specialized Department of Engineering and Machinery Entered Tokyo Tsushin Kogyo K. K.
1949	Developed the first magnetic tape in Japan
1950	Developed the first tape recorder in Japan
1951	Developed the portable recorder
1955	Developed the first transistor radio in Japan
1956	Developed the transistor television
1962	Developed the world's first broadcasting and industrial video tape recorder
1964	Assigned as Manager of SONY Specialized Machinery Team 2
1965	Developed the world's first home-use video tape recorder
1967	Assigned as Manager of SONY Development 2

Year	Biography
1969	Developed the world's first color video cassette system
1970	Assigned as Director of SONY
1974	Assigned as Executive Managing Director of SONY
1975	Developed the Betamax
1980	Developed the ultra-compact Video Movie (8mm)
1981	Developed the still camera Mavica
1982	Assigned as Senior Executive Director of SONY Developed the color video printer Mavigraph Assigned as the Director of SONY Research and Development Center
1988	Assigned as Representative Director of SONY-Kihara Research Center
1989	Assigned as General Counsel of SONY
1996	Assigned as Corporate Advisor of SONY

AWARDS

Year	Awards
1961	SONY Special Award
1967	Science and Technology Distinguished Achievement Award
1976	The 1st Ibuka Award
1977	1976 IEEE Best Paper Award
1979	1978 Eduard Rhein Award
1981	The 4th Ibuka Award
1982	1981 IEEE Best Paper Award
1982	IEEE David Sarnoff Award
1983	1982 IEEE Outstanding Paper Award
1989	Inter Camera International Honorary Award (Czechoslovakia) Photo Technology 150th Anniversary Commemorative Medal (Czechoslovakia)
1990	The Medal with Purple Ribbon
1992	University of Oklahoma Honorary Doctorate

PATENTS

Approximately 700 patents on video tape recorder etc. (approx. 330 in Japan, approx. 370 overseas)

BOOKS

"Video Tape Recorder,"

Nobutoshi Kihara, Nikkan Kogyo Shimbun

"Video Recording Technology,"

Nobutoshi Kihara, Kosaido Publishing

"Introductory Video,"

Nobutoshi Kihara et al. , Ohmsha

And more.

POSTSCRIPT

I have published many technical books on magnetic recording, but all of them used technical terms and names of parts that would be understood by engineers in the field, and I took care to precisely record how the products were made in respective times. Therefore, though they were very useful as technical references, they were not the type of books that could be understood by the general public.

This book is not a difficult technical one, but rather, an example of the technology that has underscored the history of modern technological development, and the technological history of SONY as seen from my eyes.

Still, I tended to use a lot of technical terms as I was writing, and I was having trouble trying to express this in simple language. It was thanks to the advice from Shinro Tsuchiya of Sony Magazines and editor Kaoru Toi that I was able complete this book. I am also grateful to my wife, who showed great understanding even when I forgot about my family and was lost in research and development, and enjoyed the completion of each project with me.

I would also like to express my gratitude for Mr.Akira Higuchi, Mr.Keizaburo Tozawa, Mr.Yoshio Ozaki and Mr.Shigeyuki Ochi, who willingly offered valuable historical material. And thank you to Mr.Yasuhiro Yamazaki for organizing and contributing the vast amount of photos from way back in the Totsuko era. In addition, all of the alums of the Kihara School, Akinao Horiuchi, Toshio Fujiwara, Fumio Kouno, Etsuro Saito and Katsumasa Takahashi, thank you all for your help.

Talks about publishing my technological history had crystalized within the SONY-Kihara Research Center from about a year ago, and since then, Kosuke Komatsu and my secretary Michiko Tsukamoto have gone through a great deal of trouble gathering information and proofreading the draft. I am grateful for the effort of all those involved. Thank you very much.

Lastly, I would like to dedicate this book to my unforgettable mentors and superiors, the late Mr.Shigeo Shima, Mr.Masanobu Tada, Mr.Kazuo Iwama and Mr.Michio Hatoyama.

- Nobutoshi Kihara
January 1997

ABOUT THE AUTHOR

Nobutoshi Kihara, Japanese engineer (born Oct. 14, 1926, Tokyo, Japan—died Feb. 13, 2011, Tokyo), revolutionized Sony Corp. (from 1946 to 1958 Tokyo Tsushin Kogyo K.K. [Tokyo Telecommunications Engineering Corp.]), especially with his advances in miniaturization.

He innovations that led to the creation of some 700 patents.

Working under Sony co-founder and leading engineer Masaru Ibuka, Kihara helped to develop such products as the tape recorder, the transistor radio, and the Betamax videocassette recorder.

He later became president of Sony-Kihara Research Center, which was involved in developing digital applications for image processing; Kihara remained with the company until 2006.

FIELD ARCHIVE INC.

This is a publisher launched by Tomomi Kihara, daughter of Nobutoshi
Kihara.
It is my mission to rebuild valuable books out of print all over the world as
a result of my father's out-of-print book's rebuilding.
"Predecessor's wisdom to the present power".

Field Archive Inc.

TOKYO, JAPAN

Email:info@field-archive.com

publ.field-archive.com

For information on publications

by Nobutoshi Kihara, please see our website.

www.kihara-nobutoshi.com